About Island Press

Island Press is the only nonprofit organization in the United States whose principal purpose is the publication of books on environmental issues and natural resource management. We provide solutions-oriented information to professionals, public officials, business and community leaders, and concerned citizens who are shaping responses to environmental problems.

In 2000, Island Press celebrates its sixteenth anniversary as the leading provider of timely and practical books that take a multidisciplinary approach to critical environmental concerns. Our growing list of titles reflects our commitment to bringing the best of an expanding body of literature to the environmental community throughout North America and the world.

Support for Island Press is provided by The Jenifer Altman Foundation, The Bullitt Foundation, The Mary Flagler Cary Charitable Trust, The Nathan Cummings Foundation, The Geraldine R. Dodge Foundation, The Charles Engelhard Foundation, The Ford Foundation, The German Marshall Fund of the United States, The George Gund Foundation, The Vira I. Heinz Endowment, The William and Flora Hewlett Foundation, The W. Alton Jones Foundation, The John D. and Catherine T. MacArthur Foundation, The Andrew W. Mellon Foundation, The Charles Stewart Mott Foundation, The Curtis and Edith Munson Foundation, The National Fish and Wildlife Foundation, The New-Land Foundation, The Oak Foundation, The Overbrook Foundation, The David and Lucile Packard Foundation, The Pew Charitable Trusts, The Rockefeller Brothers Fund, Rockefeller Financial Services, The Winslow Foundation, and individual donors.

About the Northern Lights Research and Education Institute

The Northern Lights Institute is a 501(c)(3) organization founded in 1981 to articulate a vision of economic, cultural, and environmental sustainability for the Northern Rockies and to provide the citizens of the region with the tools needed to achieve that vision. Since 1984, NLI has been working with the diverse citizenry, businesses, and governments of the region to reduce conflicts over the use and management of the West's natural resources and to develop equitable resource policies through the use of collaborative and consensus-based processes.

NLI has evolved a pair of programs to achieve the mission of the organization. The Community-Building Program involves active projects in the arena of natural resource and environmental conflicts, with an aim to solving problems without litigation or forms of advocacy that seem to erode more than they build community. The Research and Education Program develops written materials on western natural resource and environmental issues, with special attention paid to the evolving field of collaborative decision making.

ACROSS
THE GREAT DIVIDE

ACROSS
THE GREAT DIVIDE

EXPLORATIONS IN COLLABORATIVE
CONSERVATION AND THE AMERICAN WEST

EDITED BY PHILIP BRICK, DONALD SNOW,
AND SARAH VAN DE WETERING

ISLAND PRESS
Washington, D.C. • Covelo, California

ISLAND PRESS is a trademark of The Center for Resource Economics.

"The Quincy Library Group: A Divisive Attempt at Peace," by Ed Marston, is reprinted by permission of *High Country News* (www.hcn.org). It originally appeared as "The Timber Wars Evolve into a Divisive Attempt at Peace," *High Country News,* September 29, 1997 (vol. 29, no. 18). A number of the essays appeared originally in *Chronicle of Community;* they are reprinted by permission of *Chronicle of Community.*

Maps are by Bill Vaughn; the Heart of the Dragon illustration is by Mall Johani; woodblock prints are by Michael McCurdy.

Library of Congress Cataloging-in-Publication Data

Across the Great Divide : explorations in collaborative conservation and the American West / edited by Philip Brick, Donald Snow, and Sarah Van de Wetering.
    p. cm.
Includes index.
    ISBN 1-55963-810-9 (cloth : acid-free paper)
    — ISBN 1-55963-811-7 (paper : acid-free paper)
    1. Environmentalism—West (U.S.) 2. Environmental protection—West (U.S.)—Citizen participation. I. Brick, Philip D. II. Snow, Donald. III. Wetering, Sarah Van de.
    GE198.W47 A27 2001
    333.7'16'0978—dc21

                                              00-010480

Contents

BROADENING ENVIRONMENTAL HORIZONS   211

Foreword

In the spring of 1983, I was completing what I had decided would be my last term in the Montana legislature, and therefore my only term as Speaker of the House. I was not leaving out of any bitterness or frustration, but I did want a chance to think and maybe even write about a growing uneasiness I had come to feel over the unrelentingly adversarial way we did our work on the floor of the House . . . and as far as I could make out, everywhere else that public policy was formulated. I couldn't quite understand why Montanans, who spoke so often and so passionately about the land they loved (and seemed to love so similarly), could formulate policy for that landscape only as if they had absolutely nothing in common with one another.

I wasn't sure this was anything more than one amateur politician's very personal experience of cognitive dissonance . . . but it bothered me enough that I felt compelled to dig more deeply into it. As I cast about for some way of doing that, I came across a brand new organization, the Northern Lights Institute, which claimed to exist for the purpose of thinking harder and better about issues just like this. I got to know Northern Lights' first director, Mike Clark, and before long he had found a way to get me thinking and then writing about politics, place, and community.

In the course of our work together, Mike Clark said something that has come back to my mind repeatedly in the years since. "I've always been fascinated," Mike said, "with the way ideas move through a region." So, increasingly, have I shared that fascination, and I think I have finally learned a little something about how ideas do move. This book, in fact, seems to me an instructive episode in the movement of one very potent idea across a region. That idea is presented here under the title of "collaborative conservation."

After I left the Montana legislature and began making my way deeper into the West, I soon discovered that the need I had felt for some new approach to issues of western public policy was itself fairly widespread. Many, many people, after years of pursuing their own lives and their own best lights, had essentially worn themselves out in the trench warfare of adversarial politics and decision making. Most of them still cared enough about the places that sustained them that they would have gone on fighting in spite of the exhaustion. But now there was something new afoot—some new set of ideas moving across the landscape, offering at least the glimmer of a more promising way of doing good work, and doing it more sustainably. What it depended on, above all, was getting people whom the adversarial system had taught only to be enemies to become, if not friends, then at least constructive joint problem solvers.

The work on the ground, on behalf of land, ecosystems, watersheds, and communities, continued to be hard work, of course (and shows no sign of becoming anything else). But now there was added to it, interwoven with it, another kind of work, no less demanding, no less frustrating, no less rewarding—the work of helping those people on the ground think through the possibilities and pitfalls of a new way of doing business in the West. That's the work that people like Don Snow, Sarah Van de Wetering, Phil Brick, and most of the contributors to this book have been doing for years, work that reaches a kind of plateau in the publication of this volume—a place from which we can survey the terrain we've crossed and also see at least a little way into the forward distance.

The landscape metaphor of that last sentence is all but inescapable here, because the shape of the land and of our thought cannot be totally disentwined. Something about these western landscapes seems to demand, for example, a regionally specific kind of intellectual honesty. It's what I believe accounts, finally, for Wallace Stegner's western voice—or for A. B. Guthrie's or Terry Tempest Williams's or Bill Kittredge's or Patricia Limerick's western voices. The land demands much of its true inhabitants, not least a particular kind of truth-speaking. So when an idea or a new set of ideas comes into this landscape, or grows out of the landscape, it moves through the region on the land's own rugged terms. The ideas serve the land, and if they do not serve it, they do not survive. How do we know if they pass that test? In the end, we will know by how they prove out on the ground. In the meantime, they will be tested by the kind of intellectual rigor the region has made its hallmark.

I think the real question I carried out of the Montana legislature was whether we in the West might ever manage to create a politics worthy of this place. I still think we might. If we do, it will be in large part because of the hard, honest work of people like those who wrote and edited this book—and the honest response of the people who read it.

—DANIEL KEMMIS

Coming Home: An Introduction to Collaborative Conservation

Donald Snow

The buzz of collaboration is all around the West. You can hear it being whispered by the grasses, moaned by the trees, gurgled by the fishes, hummed by the bees. In many places, collaborative decision making seems to be emerging mostly in the form of watershed groups, but collaborative processes are also breaking out in many other settings and across nearly all environmental issues, from the reintroduction of species to the management of timber, wildlife, and grazing, to the control of suburban sprawl and the protection of valued habitats, to the abatement of industrial pollution, and more.

There is no precise, generally accepted definition of collaborative conservation, and in fact, that very term for the phenomenon we are discussing in this book is not widely used. The collaboration movement—if it is a movement—is still young enough that a standard lexicon to describe it has not yet evolved. But there is another factor beyond mere youthfulness that contributes to the amorphousness of collaborative conservation: It runs counter to the "normal" course of environmental politics, counter to the course of most politics of any kind in the United States. Collaboration tends to scramble ordinary political arrangements; it often requires a redefinition of boundaries, a crossing of borders. Indeed, that scrambling is one source of the immense power that lies latent in collaboration.

In preparing the outline for this book, my co-editors, Phil Brick and Sarah

Van de Wetering, penned one of the best descriptions I've seen of this emerging phenomenon:

> Often called "collaborative conservation," this new movement represents the new face of American conservation as we enter the twenty-first century. Although no single strategy, process, or institutional arrangement characterizes this movement, collaborative conservation emphasizes the importance of local participation, sustainable natural and human communities, inclusion of disempowered voices, and voluntary consent and compliance rather than enforcement by legal and regulatory coercion. In short, collaborative conservation reaches across the great divide connecting preservation advocates and developers, commodity producers and conservation biologists, local residents, and national interest groups to find working solutions to intractable problems that will surely languish unresolved for decades in the existing policy system.

In *Across the Great Divide*, we have gathered together some of the best available writing, reporting, and analysis of collaborative conservation. While the third part of the book, beginning on page 77, presents a series of stories about active projects in specific places, this work does not stop at the presentation of what social scientists love to call "case studies." In addition to these stories and positive examples, a critical examination of the collaboration phenomenon is presented on these pages. What are some of the social, economic, and political factors that have led to collaborative conservation? Where to collaborate and where not? What about the critical question of devolving decision-making authority to lower levels of government? How are people to find the balance between local and national authority in issues related to public lands and resources? How are the results of collaborative conservation to be evaluated? What about the use of science in collaboration? Whose voices are missing, and how can those voices be brought in and heard? These questions and others are addressed in this volume, often in the spirit of debate. There is no attempt at ultimate resolution here.

The method at work in assembling this book is the same method we at Northern Lights Institute have used in putting together the *Chronicle of Community*, the journal from which many of these essays were gleaned. Since the charter issue of the *Chronicle* in 1996, we have always billed it as a critical review of collaborative conservation and of other deliberate efforts to build community in the American West. Though Northern Lights Institute is a practitioner of collaboration, it was never our intention to have the journal act as a mere cheerleader for the collaborative approach. If this young movement is going anywhere important, it must go there with humility and critical awareness—something that all political movements ought to learn, though few, in their zeal, seem to do.

In the balance of this introductory essay, I make my own contribution to the debate and discussion collected in this book. I first attempt to trace the recent, and some of the older, roots of collaborative conservation. Next, I take a stab at describing, if not exactly defining, the general nature of these many efforts that seem to be emerging suddenly and with great force in every corner of the West. Finally, I turn to a brief, critical discussion of collaborative conservation.

Digging into the Roots

Efforts that evolved into collaborative conservation probably got their start in the arena of alternative dispute resolution (ADR), as it was applied to environmental issues beginning in the mid-1970s. ADR practitioners sought to resolve difficult environmental issues through mediation, negotiation, and the building of formal agreements among disputing parties. While environmental ADR tended to focus on resolving differences between small numbers of parties—often just two—collaborations tend to involve multiple parties, hence the metaphors "roundtables" and "stakeholder groups" used to describe collaborative entities. But many of the essential ingredients of ADR remain in place. Roundtables dedicated to collaborative conservation usually benefit from the arts and skills of effective communication developed through ADR. People accustomed to speaking in arenas of conflict on behalf of various constituencies often must learn afresh how to listen; once the listening begins, the group can begin to access its latent creativity. As ADR adherents learned, professional facilitators with no personal stake in the issue can be of immense help.

One of the fascinating features of early efforts in collaboration was the manner in which these efforts first arose (by my clock, in the years between 1985 and 1990): Numerous groups, located far apart and working mostly in isolation from one another, began meeting at about the same time, often with nothing more in mind than the need to try something new. What they wished to try was not defined by any particular *issue* common to these groups; rather, these were experiments with *process*. Could better decisions somehow be made by assembling diverse interests—interests who normally battled one another? In issues that seemed stalled, intractable, could any decisions be made at all?

Collaboration was born largely of failure, the growing recognition that lawsuits, lobbying campaigns, administrative appeals, and other straight-line approaches to hard environmental issues are often narrow, usually expensive (in more ways than one), and almost always divisive in ways that reverberate beyond the immediate issue in dispute. As many people have observed, environmental conflicts in the West tend to breed long-standing hostilities and help to fracture communities; these conflicts have been major contributors to the general sense of increasing incivility decried by many commentators. Moreover, there was the problem of gridlock: by the mid-1980s most actors in the nation's and the West's environmental debates came to realize that regardless of their political positions or the constituencies they represented, positive advancement of agendas had become stalled. The environmental movement of the 1970s had been successful enough to have engendered a highly effective opposition. As Governor William Janklow of South Dakota observed in 1986, "Anybody can stop anything now."

Early successes in collaboration began to raise awareness that in some instances, the effective resolution of long-standing issues may grow from the examination of mutual interests among competing parties. In nearly every early case, collaboration started with an urge to somehow break gridlock, to move beyond a paralyzing stalemate, to try something new. When the collaborators began to sense the power their union represented, a movement was born.

The impetus to try collaboration got a big boost, at least in the West, with the publication of Daniel Kemmis's *Community and the Politics of Place* in 1992. In his provocative book, Kemmis, a longtime political leader in Montana, exhorted his readers to perfect "the art of listening," a catchy phrase that seemed to pick up approximately where formal ADR left off. Kemmis raised the possibility that what he called "face-to-face democracy" of the kind suggested by Thomas Jefferson might be just the right tonic for western communities fractured over environmental issues.

As Kemmis made clear, the endlessness of disputes over issues that remain fundamental to the ecological and economic health of communities is virtually built into the federal system of government as it was envisioned by James Madison. Madison's was a vision of governing by "keeping people apart," keeping interest groups of all kinds engaged in a kind of perpetual warfare that no one would ever quite win. It was a government of centralized authority in which all would compete against all to tip the scales of federal favoritism. But the scales could never be allowed to tip very far or for very long. Things somehow must remain balanced.

In the West, with its massive base of federal lands and publicly owned natural resources, we have witnessed with compelling clarity the outcomes of the war of all against all. But this war is being conducted in a policy world that would have seemed alien, perhaps even bizarre, to the Founders. None of them could have envisioned the explosive growth in bureaucratic professionalism that came at the turn of the twentieth century. By the time the post–World War II "timber famines" occurred, dozens of western communities had invested their fortunes in the Progressive Era belief in "scientific management" performed by a string of benevolent government agencies empowered to act as the trustees and stewards of the nation's public lands and waterways, and to meter resource uses in ways that offer perpetual economic sustenance to local communities (two goals that often prove incompatible). When these agencies have failed, in the eyes of various segments of the public (organized increasingly into narrow-interest organizations and coalitions), the aggrieved parties have sought relief in the form of new marching orders from Congress or the administration.

Following in lockstep the Madisonian prescription, the various factions—industries, trade associations, conservation groups, economic growth councils, and many other organized interests—have spent most of the years since Earth Day 1970 engaged in battles for influence, and in recent years, those battles have often led to a frustrating stasis. Federal land and water agencies, embattled from all sides, have seen the massive erosion of public confidence in their abilities to manage natural systems. Restoration ecology, critically needed in some of the West's damaged landscapes, cannot move forward because of budgetary and procedural barriers. Key species that virtually define entire ecosystems—Columbia River basin salmon, for example—become threatened with extinction while the agencies and organizations entrusted with their care seem to stand paralyzed. The gridlock does not stop with the federal land, water, and wildlife agencies; it also extends into state and sometimes local government bodies as well. The rural West's epic resistance to land-use planning and controls against suburban sprawl appears frozen in place, as developers and environmentalists checkmate each other's policy initiatives. The war of all against all is fought on every front.

In an age of rampant cynicism about government and the arts of governing, Kemmis was bold enough to suggest that breaking the gridlock probably must involve a reawakening of the sense of a res publica, the "table" around which we all sit in a democracy, and the further possibility that through reasoned debate and discussion, we can identify and learn to obey a higher public good than the one each special interest brings to the table. But as Kemmis would be the first to acknowledge, the Jeffersonian manner of relating is almost completely alien to the Madisonian battleground we have created. For so many warriors, the battle is always just about to be won, and so the notion of finding the res publica (instead of merely dictating its location) holds little appeal.

At their best, collaboration groups seem to be trying to uncover the res publica in the great debates about the West's environment, and in doing so, they seem to be reawakening a sense of community that transcends the boundaries of narrow issues. What some of them have discovered is the great power that lies latent in the impetus to collaborate. And perhaps it is the promise of power that makes the notion of collaboration so attractive. When conservation interests can ally with other influential forces in society, the chances of having their combined goals officially ratified are raised substantially.

Some Defining Characteristics

There is perhaps no need for a precise definition of collaboration, but there are some characteristics that distinguish true collaboratives from other kinds of gatherings and coalitions.

Collaboratives are coalitions of the unalike: they are composed of people who normally do not work together or are adversaries. Often, they are efforts of last resort; they typically arise in settings and issues in which other ways of making decisions proved intractable.

Most collaboratives are power circles. Their power lies precisely in the fact that they tend to rearrange and confuse the typical polarities in natural resource issues, placing normally opposed forces in a rare position of agreement.

Virtually all collaboratives are learning circles in which participants cross-fertilize and gain from each other's expertise. This may be their most lasting value—a value that transcends the question of whether a given collaborative accomplished policy reform or measurable change.

Collaborative groups often seek innovation ahead of mere compromise. Innovation is the principal hallmark of the most successful collaboratives: the group was able to arrive at a solution together that none of the participants could have arrived at alone. It is the potential to innovate that sets many collaborations apart from other kinds of coalitions and gatherings of the like-minded. The innovations they produce are evidence of the latent creativity that seems to exist naturally in diversity. Collaboratives are creative much in the way cities are creative: composed of many different parts, they can be assembled into highly constructive entities—if the right kind of leadership is present.

Like many small groups and organizations, collaboratives are often clear reflections of their own leadership, the personalities of their leaders; they tend to use patterns of mediative leadership instead of polarizing leadership. Mediative leadership seeks to bring disparate people and interests together to combine knowledge, skills, and power in pursuit of a common goal. Polarizing leadership tends to insist that "our side can win" if we can create a critical mass of support. Polarizing leadership tends to name enemies; mediative leadership tends to embrace enemies.

Collaborative groups are usually ad hoc and ex parte. They often lack corporate status and are informal in structure. Their informal nature is often a key feature of their effectiveness: informality adds to the sense that involvement in a given collaborative is voluntary and experimental. Tentative members of the group thus feel that they can opt out at any time and that decisions are not apt to be binding. These hallmark features of informality lend an air of flexibility and experimentation to the effort, at least in the beginning, when attendance at meetings may be a tenuous venture for many members.

Most collaboratives are nongovernmental in origin; they may include agents of government, but their origin typically lies outside of the agencies. As more governmental agencies enter into the field of collaborative conservation, and as they are given mandates to "go forth and collaborate," this essential feature of the movement perhaps will change. But the change will probably be for the worse: nongovernmental origin seems so far to be one of the strongest features of collaborative problem solving.

Collaboratives deal explicitly with questions of process, especially the decision-making process within the group. Many collaboratives are true consensus groups, in that they use consensus processes—often leading to formal, written agreements—to make decisions and bind the group behind common action. But too many commentators use the terms "consensus" and "consensus group" far too loosely: many collaborative groups do not use consensus processes at all. (Matthew McKinney's essay, on page 33, is especially insightful on this point.)

Much has been made of the notion that collaborative conservation is "place-based," but that's not always true. Some notably successful collaborations have operated in the abstract realm of policy. For example, the 1994–95 effort in Montana to create a statewide water leasing program for instream flows was a classic collaboration, and entirely placeless ("statewide" is not a place). Still, there is a strong sense of place-centeredness to this movement. As Luther Propst and Susan Culp point out in the essay on page 213, there is a genius loci, or genius of place, that seems to both attract and strengthen efforts in collaboration.

Finally, collaboratives are often, but not always, local or regional in terms of their scope of political sovereignty. There are many attempts at collaboration on the national level, too. That they are always local is, again, one of the myths about collaboration.

A Few Implications, a Few Red Herrings

Collaborative conservation, to me, implies three things:

1. The deep involvement of communities in the conservation and care of nearby natural resources, for the benefit of people and nature together.

2. The conservation of community itself—of the attitudes, processes, duties, responsibilities and relationships that go into forming and maintaining healthy communities, wherever they occur.
3. A sense of community involvement that is broader than local: a community of all who share a passion for accessible nature—for those settings and resources of the nonhuman world that add such great value to the experience of living on Earth.

The third sense of the term collaborative conservation is especially important, for it goes to the root of legitimate concerns about the management and care of public resources, which are, and should remain, national and even international in scope and effect. I'll return to this last point in a moment.

For many in the West—including many environmentalists—collaborative conservation represents a kind of homecoming, a way of bringing the implementation of sound environmental policy down to the ground and back into the lives of people who are directly affected by the outcomes. As a practicing environmentalist since the mid-1970s, I have long been concerned that far too much of our work has tended to alienate the very people who can cause good conservation to happen, or who can block it through inaction, the never-ending search for loopholes, or just plain recalcitrance. So much environmental policy is coercive in its effects, and so few westerners—especially those on the land—are apt to sit still for coercion. Collaborative conservation of the kind I envision fosters on-the-ground responsibility by sharing both accountability and power.

Where public lands management intersects with the economic health of local communities—a condition that applies to most of the rural West if not the urban—the notion of sharing power (or, in today's popular language "devolving power to local levels") is understandably troubling to many. And here we come back to the difficult question of the extralocal, often expressed as the national interest in our public resources.

In the May 13, 1996, issue of *High Country News,* Sierra Club chairman Michael McCloskey pointed out that in some instances, local stakeholder groups working on public lands issues represent a redistribution of power. McClosky worried openly about the implications: "This redistribution of power," he wrote, "is designed to disempower our constituency, which is heavily urban. Few urbanites are represented as stakeholders in communities surrounding national forests. Few of the proposals for stakeholder involvement provide any way for distant stakeholders to be effectively represented."

Clearly, these are wise and legitimate concerns of the kind that demand discussion and debate as collaborative decision making continues to unfold. Myopia abounds everywhere, but nowhere is shortsightedness more apparent than among local boosters of the kind who seem to proliferate in the small towns and cities of the West. Many environmentalists remain concerned that any loss or significant reduction of federal management authority will inevitably lower the possibilities for a long-range, conservationist vision to guide the management and care of the West's public lands.

But it's simply too easy to leap to the conclusion that collaborative conservation automatically translates to the triumph of shortsightedness; and easier still to assume that federal land managers (or any agents of centralized government) will look out for the public interest because legislative mandates or the courts tell them that they must. If large bureaucracies are skilled in anything, they are surely experts in the arts of inertia. The "old" conservation paradigm relies heavily on the delegation of power and managerial responsibility exclusively to agencies of government; under this paradigm, the agencies are often mandated to "accept public commentary" and to then make decisions "behind the scenes." Close observers are often left to wonder what effect, if any, the commentary had on the decisions. The "new" paradigm of collaborative conservation seeks to formalize the recognition that agencies must be responsive to local interests in ways that reach well beyond time-honored tendencies to work in collusion with private business. With agencies "at the table," decisions are often more transparent, and more clearly reflections of stakeholder involvement.

What some environmentalists seem to overlook when they reflexively attack the growing use of local collaborative conservation groups is that it was the very work of environmentalists themselves that opened the door to these kinds of new experiments with accountability and power. Having demanded public involvement for more than twenty-five years, environmentalists are now seeing one of the healthy outcomes of that demand: local collaboration groups, which now seek to implement what environmental laws have tried to mandate for a generation.

Clearly, national interests ought to be well represented in any groups that seek to influence public lands decisions, no matter where those groups locate themselves. Indeed, one of the limiting factors in using collaboration groups to improve public lands management is the scarcity of potent, experienced environmental activists in some of the most rural reaches of the West. Without those voices at the table, local collaboratives can easily devolve into cheerleading squads for the status quo, whose main purpose is to help government agencies and/or corporate interests proceed even more efficiently with predetermined agendas to "develop" public lands in the name of "local economic diversification" or some other unquestionable holy. Fertile conflict lurks around the edges of every successful collaborative I've seen or studied—and that's a good thing. Where the edge of conflict can be maintained, there is probably reasonable vigilance against the forms of collusion that have damaged the landscapes and waterways of the West. Fertile skepticism keeps fertile conflict alive.

This leads us to the question of power and powerlessness. A good share of many objections to the collaborative approach seems to issue from the belief that citizens (no matter how well organized) are virtually powerless, while corporations or agencies hold essentially all the keys to power. Yet in order for a collaboration to be true to the public interest, and to succeed, there has to be a roughly equal power equation among the stakeholders, within the context of the issues at hand. This does not imply that there must be an equality of magnitude among the parties at the table, for such parity almost never exists. Large corporations, large government bureaucracies will almost invariably dwarf the nongovernmental organizations and other small groups represented at the typical collaboration table. Nevertheless, there are many issues in which environmentalists, and other citizen groups, despite their scarce resources, hold very significant powers—endangered species designations, for example. It is the very strength of environmental laws and regulatory structures that has provided the impetus to negotiate in many issues. As long as environmental organizations remain powerful players under law, their stake in the issues cannot merely be brushed aside.

This is a critical concern among many individuals and organizations who eschew collaboration. They understand that there are also many resource issues and settings in which the power equation a priori is tipped so vastly out of balance that there will never be parity until the law is changed. Despite much recent right-wing ranting to the contrary, environmentalists are usually the ones who bear the brunt of disadvantage. Perhaps the most outrageous example comes with federal hard rock mining law. Industry claims a right to mine on federal land. Stopping the granting of a mining permit on Bureau of Land Management or national forest domain is extremely difficult. When the mining industry asks, then, for collaboration over a given mine permitting proposal, environmentalists would be well within bounds to refuse to participate, since they know the game is rigged by statute: permission to mine is often nonnegotiable. All that may be negotiated are "mitigating measures." In such instances, negotiation probably implies a battle lost before it is begun.

Finally, the question of representation at the table comes up repeatedly in most discussion about collaborative conservation. The easy bromide, accepted by many, is that all affected parties must be represented. I disagree. A better, more realistic way to envision optimum representation is this: Interests attached to all appropriate levels of sovereign authority in the issues under discussion need to be present and well represented. In this odd phrasing I am trying to avoid the easy bromide—partly because complete representation is often a practical impossibility, and partly because it's a ludicrous standard we have come to demand of collaborative conservation when we never demanded it before.

When I worked as a paid environmental lobbyist, I tried to shotgun all manner of issues through various legislatures and agencies of government in spite of any other representation, except constituencies that mattered because of either politics or convenience. Suddenly now, howls of protest over inadequate representation come roaring across the landscape when a community collaborative dares to concoct a policy proposal without checking in with every conceivable interest from sea to shining sea. There is a double standard here.

Clearly, a collaborative group is apt to get a better result, and a more palatable result, if it is not just broadly inclusive (perhaps in honesty the best that can be expected), but more specifically, if it includes interests that attach to all appropriate levels of sovereign authority. The National Audubon Society and National Wildlife Federation, for example, may occupy special positions of influence and interest in any given federal lands issue of interest to a community collaborative, and may have staff or appropriately empowered members near enough to attend meetings of the group. It would seem fruitless to proceed without them. But there are dozens of instances where the arenas of sovereignty would not demand national representation, and perhaps some arenas where local representation would be unnecessary.

Collaborative conservation of the kind I'm describing demands the practice of old-fashioned citizenship and a loosening of the grip on narrow-interest politics for all involved. It is not performed well by the ideologically pristine or by those whose positions are secured by their refusal to join in the responsibilities of governing. My reading of the West's political history tells me that collaborative conservation will have a tough battle in order to survive, for the West is not a region with a long, robust history of accepting political responsibility. The federal presence and the public lands themselves (though they are indeed political lands) have helped insulate the West from the kind of local responsibility demanded by community-based collaborative conservation. But the gauntlet is now down, and the burden of governance must be shared by all who deign to try it.

FROM TROUBLED WATERS: THE EMERGENCE OF COLLABORATIVE CONSERVATION

Perhaps reflecting a more general trend in American politics in the 1980s and 1990s, environmental politics have become increasingly polarized and polarizing in recent years. Although conflicts over natural resources have always been contentious, we now refer to many of them, especially disputes about the fate of resources on public lands in the American West, by making ample use of war metaphors. Despite the escalating rhetoric, existing processes and institutions for resolving environmental conflict seem more poorly equipped than ever to handle such conflicts. The result is policy gridlock, which only enhances a sense of frustration and alienation for everyone.

In response to this sense of frustration, American environmentalism is in the throes of a transformation that rivals the movement's shift from protest politics in the 1960s to its institutionalization in American politics in the 1970s and 1980s. By 1990, at the same time that environmental groups seemed to be at the pinnacle of their power, green groups also ran into increasingly fierce opposition from farmers, ranchers, miners, loggers, and powerful conservative ideologues, forcing important environmental decisions into political gridlock. But amid the policy gridlock that now characterizes most environmental debates, including saving remnant ancient forests, restoring riparian habitats for native and anadromous fish, reforming outdated mining laws, and protecting endangered species and ecosystems, a new conservation movement has emerged, emphasizing the importance of consent, not power, in environmental strategy. We call this new movement collaborative conservation, which encompasses a wide variety of local partnerships, library groups, resource advisory councils, river and range conservation initiatives, and watershed councils.

This part describes the policy context in which these collaborative groups have arisen. The first two essays address questions such as these: Why have traditional environmental approaches reached a dead end, opening the door to more collaborative approaches? How well will collaborative efforts mix with more traditional approaches? What does the collaborative conservation movement add to broader environmental goals and to the environmental movement itself? The final essay, by Matthew McKinney, introduces us to the idea of consensus, an integral concept in the collaborative conservation movement, even though not all collaboratives can or should operate by consensus. Similarly, although there is no how-to guide to successful collaboratives, McKinney's essay does point us in the direction of some key ingredients that have been central to successful efforts so far.

Will Rain Follow the Plow? Unearthing a New Environmental Movement

Philip Brick and Edward P. Weber

In the wide-open spaces of the American West, where visitors can step out of their cars and almost feel themselves expand to meet distant horizons, a new environmental movement is emerging from the soil. This movement, which we call collaborative conservation, is a movement premised on the hope that we can expand our notions of individual and community to meet ecological and social challenges as wide as the western skies themselves. Collaborative conservation is a young movement brimming with optimism, much like early settlers to the region who believed that hard work and good intentions alone would transform the region's arid climate—that rain would follow the plow. But can this new environmental movement stake out new ground? This will be difficult, because although collaborative conservation is a qualitatively new approach to protecting natural resources, it is also an approach riddled with paradox. Collaboratives are a response to dysfunctional environmental strategies and policy processes, but are also symbiotically dependent on them.

Part of the charisma of collaborative groups is that they produce political positions that are not easily placed in traditional boxes. As interest group theorists Tierney and Frasure describe these boxes, "the preservationists generally espouse communitarian values, active government, and the ideals of a 'postmaterial' culture; the users tend to be advocates of private property rights, personal autonomy, individualism, and limited government." But what happens when "preservationists" pursue their postmaterial values in the context of property rights and limited but enlightened government? When advocates of limited government pursue postmaterial and communitarian values? Powerful coalitions for ecological goals can emerge. And it's not as if local collaborative groups have a monopoly on crossover constituencies and strategies: two of the most successful and prominent national environmental groups, The Nature Conservancy and Environmental Defense, occupy this crossover political ground.

But stepping outside traditional boxes puts collaboratives in a difficult position: no one knows which box to put them in. It is certainly possible to attend both wise use and environmental gatherings where nearly identical expletives are used to describe collaborative groups. This no doubt reinforces the self-perception of many collaborative participants as occupying a political space that can be described only in ironic terms: the radical center.

A Furrow Deep and Wide: Contours of a New Environmental Movement

At this radical center is collaborative conservation, a movement that has the potential to redefine the language of environmentalism. It is not the preservationism of John Muir, or the traditional conservationism of Gifford Pinchot, nor is it the top-down environmentalism that characterizes most national environmental groups' approach to natural resource policy. Emerging in the late 1980s and early 1990s, collaborative conservation advocates emphasize community building, guided by a larger mission embracing ecological and economic sustainability. Some collaboratives seek to devolve authority away from federal and state land managers to more local, place-based citizen networks. Others are content with federal authority, but want more meaningful citizen participation. In either case, the concept of "place" is a central organizing principle. These places are usually rural communities intimately connected with nature's bounty, such as ranching, forestry, agriculture, commercial fisheries, outdoor recreation, and tourism communities. The biophysical scale of place varies widely among groups, and is almost never determined scientifically, as some proponents of ecosystem thinking have suggested. Instead, place is more an intuitive, intersubjective sense of what constitutes "here," as opposed to what is more "over there," though in the West these intuitions are often quite clearly shaped by the surrounding topography—valleys, mountains, watersheds, and the like. In short, place is both a physical and political space, with powerful implications: Place is almost universally understood by collaborative advocates to be the foundation and catalyst for enlightened self-governance, despite the differing interests that loggers, ranchers, environmentalists, Native Americans, kayakers, hunting guides, county officials, land managers, and other interested citizens bring to the table.

The emphasis on place is not the only feature distinguishing the collaborative conservation movement from its preservation, traditional conservation, and contemporary environmental brethren. At least three further distinctions are useful: Where contemporary environmentalism emphasizes ecocentrism, collaborative conservation integrates ecocentric and anthropocentric goals; where most environmentalists embrace regulatory democracy, collaboratives prefer civic democracy; and where environmentalists put great faith in science and technocratic management, collaboration advocates seek to integrate science with local knowledge. In many ways, collaborative conservation captures many of the early ideals of the contemporary environmental movement (decentralization, renewal of community), and joins them with a creative blend of preservation and traditional conservation.

Balancing Ecocentrism and Anthropocentrism

Since the late 1980s, the environmental movement has become increasingly eco-centric in orientation, pursuing ever more aggressive measures to protect and isolate nature from the ravages of industrial civilization. Policy proposals increasingly emphasize the importance of nature *over* humanity. In some circles, nature not only has intrinsic worth apart from humans, but it also has rights on par with humans. Policy goals emphasize severely restricting human impacts on nature, as illustrated by the zero-discharge goal of the Clean Air Act of 1972, and the prohibition on considering cost when determining if a species qualifies for protection under the Endangered Species Act of 1973. In short, these environmentalists believe that nature must always be the winner in any serious conflict with economic goals, even if these goals are fundamentally redistributive.

Collaboratives, on the other hand, seek to meld nature together with economy and community, and see the separation of humans from nature as an impossible task given the demands and desires of a growing population. Rather than viewing humans and nature in a zero-sum struggle, the new collaborative movement hopes to meet ecological and human needs simultaneously. For example, when ecological goals might produce economic dislocation for some community members, planning for sustainable, family-wage job alternatives is integrated into the discussion of the ecological goal in the first place, not as an afterthought, as in the spotted owl controversy in the Pacific Northwest.

From Regulatory Democracy to Civic Democracy

Here is where the collaborative conservation movement breaks most dramatically with its predecessors, challenging fundamental assumptions about the nature of the individual, the market, and the scope of government. Environmentalists have long assumed that given the liberal individualism that pervades American culture, government is the primary, if not the only, institution capable of defending nature against inevitable market failures. Preservationists from Muir forward have called on government to insulate nature from industrialism. Traditional conservationists hoped to combine expertise with state power to manage resources for the benefit of all. But historically, traditional conservation is more closely associated with cozy relations between private interests and public institutions in pursuit of economic growth. Contemporary environmentalism, on the other hand, also views government intervention as a necessary corrective to market failure to protect nature, but differs from traditional conservation by trying to create a sharp dividing line between public institutions and private interests. Ideally, and perhaps ironically, environmentalists hope to replace the cozy relations between private industry and federal land managers with equally cozy relations of their own.

Today's environmentalists hope to effect environmental goals on a national scale, emphasizing compliance with federally mandated environmental standards and review procedures. Those out of compliance with the standards and procedures are sued, often successfully. In this sense, environmental power is best expressed as the ability to force compliance on unwilling or recalcitrant subjects.

Collaborative advocates think differently. Instead of viewing individuals in the classic liberal sense, individuals are better understood in the context of the communities they inhabit. Self-interested individuals still exist, but are profoundly shaped by social interaction with others. Preserving nature is much more than adopting and enforcing a set of formal, legal rules governing interaction with nature. Instead, collaborative conservationists emphasize participation in civic institutions, a political sphere somewhere between government and market. In this context, collaborative advocates are attempting to carve out a space for autonomous action within the larger top-down system of governance that is the legacy of the three preceding movements. Instead of a system premised on hierarchy, collaboratives devolve significant authority to citizens, with an emphasis on voluntary participation and compliance, unleashing untapped potential for innovation latent in any regulated environment. It also seeks to mediate national environmental law by balancing it against local ecological, economic, and social conditions. And despite criticism by leaders of national environmental groups, most collaborative conservation participants understand themselves to be working within, not against, a larger framework of national environmental law.

From Technocracy to Integrated Knowledge

Another striking difference between collaborative conservation and its predecessors is its emphasis on integrating local and scientific expertise. Environmentalists have always appreciated the importance of scientific and technical expertise, and this appreciation has grown as more environmental groups gain expertise and professionalism of their own. The rapidly expanding fields of ecology and conservation biology are only adding fuel to this fire, both by widening the scope of ecological concerns and by creating new cadres of experts to address these concerns. These experts are supposed to be insulated from the corrupting influences of politics, free to make natural resource policy decisions based on scientific information alone, and this scientific information is expected to apply equally well across a variety of landscapes.

Collaboratives embrace the importance of scientific knowledge and expertise, but at the same time seek to expand the concept of expertise beyond bureaucratic and organized interest expertise. Collaboratives are also less certain that scientific information can or should be insulated from the political arbitration of conflicting values. Explicit attempts are made to integrate scientific knowledge and technical expertise with local knowledge, a community perspective, and local talents. Citizen concerns about quality of life, local conservation practices and stories, and folk knowledge, for example, are treated on par with recommendations made by scientifically trained experts. Again, this approach blurs the distinction between public and private, so important to the traditional conservation and environmental perspectives. But at the same time, it opens up traditional science to new ways of knowing, and perhaps new understandings about how complex ecosystems function. It also creates a greater sense of community investment in efforts to protect nature, as it validates the importance of local stories and talents, which rarely command the prestige of knowledge produced at remote universities and research centers.

Although there are certainly more subtle differences between collaborative conservation and earlier movements, the emphasis on integrated knowledge, civic participation, and ecological holism clearly distinguishes collaboratives from their environmental predecessors. Not coincidentally, these three features correspond to three strategic weaknesses that plague contemporary environmentalists.

A Wheel in the Ditch, and a Wheel on the Track

Environmental power, based on a phalanx of environmental laws, administrative procedures, grassroots activism, and political muscle in Washington, D.C., and in state capitols, has grown dramatically since 1970. There is little doubt that this growing power has produced important achievements, but at the same time, the limits of environmentalism's power became increasingly evident in the 1990s, as the regulatory strategies pursued by many environmental groups seemed out of sync with increasingly conservative trends in national politics. What happened?

The End of Social Regulation

For better or for worse, Americans are increasingly suspicious of government's ability to intervene in economic and social affairs. At the same time, however, we also seem to want environmental amenities that private action alone cannot guarantee. For example, when our love for nature contributes to the sprawl we deplore, it is tempting to look for convenient legal weapons to prevent it. A favorite environmental weapon is the Endangered Species Act. For example, the Center for Biological Diversity in Arizona is using the cactus ferruginous pygmy owl to halt sprawl in the greater Tucson area. This is asking a pocket-sized owl to shoulder a tremendous burden, threatening political support for the act. As Fred L. Smith, president of the Competitive Enterprise Institute, predicts, "over the next several years, the environmental laws are going to go the way of welfare laws. Instead of welfare waivers, states are going to be asking for the equivalent of ecological waivers." If this is indeed the case, it spells trouble for traditional environmental approaches, which rest on the power of existing environmental law. But at the same time, devolution will create many opportunities for creative collaborations.

Many environmentalists are skeptical about local collaboratives, charging that collaboratives naively ignore the importance of power and coercion in solving natural resource conflicts. In the environmentalists' view, natural resource producers, land speculators, and developers have all the power, so why sit down and collaborate from a position of weakness? Although there is some truth to this concern, it ignores two important facts. First, relying on the coercive power of government is less "powerful" than most environmentalists think. Political scientists have long recognized that the greater the coercion and compliance costs implied by a regulation, the less likely it will be enacted. And if enacted, such regulation is more likely to be applied and enforced symbolically than substantively.

Second, critics of collaboration often forget that raw coercion is not the only form of power. A more potent understanding of power, following Hannah Arendt, is the human ability to act in concert with others who share similar concerns. In this sense, power is more about mutual enablement than about domination of one by another. By their very nature, collaboration conservation efforts have the most potential to develop this form of power and add it to the environmental arsenal. As the Sierra Club and Wilderness Society rather rudely discovered in congressional hearings on the Quincy Library Group (see page 79), the idea of former adversaries working in concert with each other is politically irresistible.

Just Another Interest Group

To be sure, the success of environmental groups in becoming part of the power establishment, both formally and informally, represents an important accomplishment, even though this opens the movement up to legitimate criticism from its more radically inclined fringes. It also opens the door to a more serious problem. Once in the company of insiders, it becomes increasingly difficult to distinguish oneself from the myriad other interests seeking political favors. As political theorist Benjamin Barber writes, "citizen groups organized in desperate defense of the public interest find themselves cast as mere exemplars of plundering private interest lobbies." Barber predicts this problem will only get worse as the power of global capital becomes ever more pervasive.

Environmentalists have indeed noticed this problem, and have begun to try different strategies. According to Paul Gilding, the former head of Greenpeace, "the old model is you go into the streets, get the government to pass a law, the companies resist, eventually a compromise is reached and in five years things change." "But the smart activists," Mr. Gilding says, "understand that the market, and global integration, is now king," and that the groups making the biggest difference today are those working with multinationals and consumers, showing how they can be both green and profitable. Those companies that refuse to talk will be the target of global consumer boycotts organized over the Internet, so there is a great deal of incentive for industry to clean up its act.

At first glance, this strategy looks promising. It contains elements of both engagement and coercion, two ingredients absolutely necessary for any successful collaborative enterprise, however paradoxical that may seem. It looks like a classic "win-win" strategy. However, it seems utterly inconceivable that we can expect to make a meaningful dent in our common ecological woes if we reduce our concept of effective political action to a more selective breed of consumerism, which by definition limits our field of vision. What about citizenship, which requires moving beyond mere self-interest to visualize the common good? Before we rush headlong to embrace the market-based strategies proposed by Mr. Gilding, environmentalists should be careful not to overlook the potential latent in local civic communities.

Ironically, as the forces of globalization weaken sovereignty at the national level, these forces are awakening interest in more local sovereignties, dormant for more than a century while we believed that the best road to prosperity was a national economy. Now that a global economy seems inevitable, interest in the integrity of local communities is rising, and no one expects the federal government to consistently intervene in the global, free-trade megamachine on behalf of local interests. If citizens want environmental amenities, which can only be produced by acting in concert with others, then local collaborative initiatives may be the best available means to achieve them.

Urban-Rural Divisions: Environmentalists Are from D.C., Wise Users Are from Mars

Rural landscapes, both pastoral and wild, run deep in the American psyche. We revere our countryside, even though most of us live in cities. But at the same time, do we revere those who live and work in the countryside? Contemporary environmentalists seem to have particular difficulty with this question, and answers always seem to lapse into a shopworn debate about jobs versus the environment, which doesn't do justice to the complexities of either.

But in poll after poll, westerners in rural areas consistently demonstrate high levels of ecological awareness and concern, but until recently, have not been able to articulate these concerns in the context of prevailing environmental discourses and policy initiatives. Instead, many rural westerners have found their voice in the wise-use movement, which, among other things, vows to destroy the environmental movement as currently constituted. This is mistakenly taken as a sign that rural westerners are not worried about the economic, social, and environmental changes that are transforming their communities. Nothing could be further from the truth. The very possibility that a coalition of antienvironmental interests could steal the term "wise use," once an organizing principle of the American conservation movement, should signal that the environmental movement has lost its footing in rural areas. In their justifiable zeal to protect nature, many environmentalists seem to have forgotten that we are all implicated in the environmental crisis. We all use resources.

There are still deeper dimensions to the urban-rural divide in the context of contemporary environmental strategy. Two problems come readily to mind, both linked to how environmentalists relate to work. First, many of the legal-administrative strategies used by environmentalists systematically exclude, however unintentionally, the interests and concerns of rural labor. In the current policy system, achieving environmental goals requires participation in complex and bureaucratic processes, as well as expertise in legal and scientific discourses. Has the environmental movement helped equip rural

constituencies with such skills? Or are local preferences for informal procedures, commitments confirmed by a handshake, and indigenous knowledge
simply dismissed as retro, redneck, or random?

Second, the past three decades of environmental activism have, by and
large, alienated the labor movement. Yet both are social movements focused
on social justice and better living conditions. So why are environmentalists
often held in low esteem in working-class circles when polls show a deep populist concern for environmental quality?

A bumper sticker often seen in rural areas of the West reads, "Are You an
Environmentalist, or Do You Work for a Living?" This public characterization
of environmentalism poses a serious political problem for environmentalists.
Many of the environmental movement's goals, at least in the public lands battles in the West, have more to do with play than with work. Environmentalists don't propose to work in wilderness areas, they want to play there. Worse,
as historian Richard White writes, "environmentalists have come to associate
work—particularly blue collar work—with environmental degradation. This
is true whether the work is in the woods, on the sea, in a refinery, in a pulp
mill, or in a farmer's field or a rancher's pasture. . . . Nature seems safest when
shielded from human labor." What alienates workers from environmentalists
is that the latter seem unwilling to acknowledge the ecological implications of
their own work. Most members of mainstream environmental groups come
from occupations that manipulate symbols, not things, which in their view is
somehow less destructive to the environment. But as a favorite rural apho-
rism goes, "A single lawyer or accountant can, on a good day, put Paul Bun-
yan to shame."

The disconnect between labor and environmentalists, however, may be
about to change. The last big environmental rally of the twentieth century,
at the World Trade Organization (WTO) protests in Seattle, may signal a
new, antiglobal populism that integrates labor and environmental concerns. A labor-environmental coalition at the national level to fight the
excesses of a global free-trading system is welcome, but will this new coalition translate into better labor-environmental relations in rural communities across the West? If there is any place for such alliances to be (re)dis-
covered, it is most likely to emerge in the processes of collaborative
dialogue, not in administrative appeals or in symbolic, media-centered
shouting matches.

A New Rural Populism?

Ultimately, the fate of collaborative conservation efforts will be tied to their
ability to cultivate a home for newly emerging political identities in an
increasingly fluid political environment. Collaboration's power lies in its

promise of a new rural populism, an irresistible alliance of traditional political opposites, molded into novel civic- and ecologically minded identities.

For many of us, thinking about a new rural populism requires us to think beyond the insular rhetoric, paranoia, and ethnic recriminations that characterized both the 1890s and 1990s populisms that spread from America's heartland. But western political, social, economic, and demographic landscapes are changing faster than the categories we invent to understand them. Even though the violent, rural populism of Timothy McVeigh and the militia movement has been better at grabbing headlines, the quiet work of collaborative conservation groups demonstrates that a far more potent and inclusive brand of populism can emerge from creative rural communities. In the West's new landscape, a host of new political orientations is in the offing. For example, so-called lifestyle conservatives, who are moving to the interior West in epic numbers, have a suspicion of government that is matched only by their concern about the erosion of the communities and environment that attracted them to the region in the first place. Would these folks have a lot in common with the more traditionally liberal, suburban, Sierra Club member? Similarly, fiscal conservatives serious about slashing government waste might find new friends in the Wilderness Society, which opposes federally subsidized resource extraction from public lands. The possibilities for collaboration and new political identities are endless, raising tantalizing possibilities for a new breed of American environmental populism, just in time to meet new challenges posed by globalization. It may be too late to stop the globalization train, but won't it be interesting to be a fly on the wall when an Earth First!er discovers that she or he has a lot in common with an evangelical Christian talking about the importance of saving all the parts, just like Noah and his ark?

REFERENCES

Arendt, Hannah, *On Violence*. New York: Harcourt Brace, 1969.

Barber, Benjamin, *Jihad vs. McWorld*. New York: Ballantine, 1995.

White, Richard, "'Are You an Environmentalist or Do You Work for a Living': Work and Nature," in William Cronon, ed., *Uncommon Ground*. New York: Norton, 1995, 172.

Tierney, John, and William Frasure, "Culture Wars on the Frontier: Interests, Values, and Policy Narratives in Public Lands Politics," in Allan J. Cigler and Burdett A. Loomis, eds., *Interest Group Politics*. Washington, D.C.: Congressional Quarterly Press, 1998.

ONRC, Go Home: A Rancher Speaks Out to Environmentalists about Community and the Land

Mike Connelly

You know, I never thought I would ever say this, but I'm starting to miss Andy Kerr. Ever since he left, it seems like the Oregon Natural Resources Council (ONRC), which he helped build into one of the most effective environmental organizations around, has suffered from what seems like a crippling lack of vision. In Klamath County, Oregon, where I live, the two most recent ONRC lawsuits are very disturbing to me, not so much because their stated purpose is to put many of my friends and colleagues out of business, but because they really seem to be just plain disruptive for disruption's sake.

With Andy, you always had the sense that ONRC actions were guided by very concrete and well-defined goals, that there was always some shining prize they always kept their eyes on. Anymore, it seems as if the regional representatives just sit back, watching other stakeholders participate in admittedly tedious efforts to find equitable solutions to very complicated problems. And then, when the conclusions do not precisely coincide with ONRC demands, they file a lawsuit, claiming that they gave everybody every opportunity to submit to the pristine and irrefutable truth of ONRC positions. "What else could we do?" they ask innocently. "They just wouldn't listen to reason."

Random litigation may be a tried-and-true means of publicizing one's advocacy group, but in the early days the ONRC did more than raise money.

It inspired people, right or wrong, to make great sacrifices in the name of realizing a vision. Many have noted that for the most part, the source of this vision was Andy Kerr, and I think we all recall the palpable panic among sympathetic environmentalists as Andy announced his resignation. I think the main reason I miss Andy is that he possessed a kind of heedless honesty, a willingness to face head-on and sort out contradictions within his own constituency, and not just those of his enemies.

I was reminded of this recently as I read a short piece Andy wrote about cowboy music. It was called "'Home on the Range' is Actually an Environmental Folk Song," and in it he confesses to the "inconsistency" of his liking cowboy music, even though he "can take or leave cowboys." He quotes the lyrics to "Home on the Range" in their entirety, and makes the observation that this "most well-known of cowboy folk songs does not contain a single reference to cows."

His point, of course, is that livestock production is not central to the appeal of western lifestyles and landscapes, and that eliminating livestock from the range would do nothing but enhance those characteristics that cowboys "really" valued. He ends the piece by suggesting a new verse:

> Oh it will not be long 'til the livestock are gone,
>
> And the bighorn range without fear;
>
> When the native biotic will retake the exotic,
>
> And the streams again will run clear.

But Andy, you came so close to getting it!

No, the song doesn't mention cows, but as Andy himself points out, along with "wide open spaces, clean air, bright stars, birds and flowers," cowboy music is primarily concerned with "friendship, freedom, love, honor and duty." I would add family and history to that list, and then remind Mr. Kerr that the name of the song is "HOME on the Range."

Now, when I think of my "home"—and I bet this is true of most folks—I don't think of just landscapes or just animals or just occupations or just family and friends. My memories of my childhood home, not to mention my affections for my present one, are made up of an inseparable mixture of all of these things, a mixture that always comes out of my head in the form of stories—stories about certain people doing a particular set of things in very specific landscapes. Like picking grapes with Mexican "gangas" as the early morning mists slid under Stag's Leap. Or the time the whole neighborhood turned out to chase a maniacal steer through farm and field, only to finally watch it shot and slaughtered right there on the manicured lawn of a local millionaire.

I suppose I could write a song about these things, too. And while it may never achieve "folk" status, my song would be doing the same thing for me that "Home on the Range" did for long-ago cowboys: wrap up all my experiences, sensory and otherwise, and stick them in the ground somewhere down there where I was raised, or where I am right now—down home.

What I'm talking about here—this elusive soup of landscape and community, of people and place, all kept together with a collection of shared stories—is not the sole property of cowboys, or even of agriculture. In fact, most of the best talk about these matters has come out of the environmental movement under the heading of "bioregionalism" or "sense of place."

Which brings me back to the ONRC, since my guess is that the ONRC has reached a point in its history where it might do it some good to start shifting at least part of its efforts toward working out the real, live nuts and bolts of how human communities are supposed to draw sustenance from their respective homes. Ever since the emphasis has shifted from corporate logging and the spotted owl to farming and ranching, the ONRC seems to be stumbling over its inability to conjure up any corporate demons or alien behemoths to serve as targets for urban-environmental rage.

In Klamath County, where there simply isn't any corporate farming or ranching to speak of, the ONRC has the monumental task of trying to convince potential supporters that a family like mine—with two little kids, living off around $15,000 a year—is the fire-breathing incarnation of industrial agriculture, responsible for heartlessly sacrificing birds and fish in the never-ending quest to satisfy our infinite and unmitigated greed. Klamath County agriculture really is just a bunch of family farms, and to its credit, even the ONRC seems to be realizing that the tactics that got this group so far on the spotted owl issue may not work so well when the only boutique species you have are cryptobiotic crust and bottom feeders.

The ONRC has recently announced the "Oregon Wild Campaign," through which it hopes to reclaim some of the vitality that characterized the organization back even before the spotted owl came along by focusing on federal designation of wilderness areas. The public outreach segment of this campaign is the "Adopt a Wilderness" program, whereby individual supporters become "parents of place" by helping with the assessment, mapping, interpretation, monitoring, and advocacy for a particular tract of land.

The point I'd like to make about this program is that, like urban American environmentalism in general, it seems to me to be rooted more in the institutionalized separation of humans and nature than in an acknowledgment of interconnectedness and interdependence, more in unwitting assertions of human superiority than in moves toward ecological humility.

The ONRC's promotional literature for the Oregon Wild Campaign makes use of three common metaphors in attempting to justify federal wilderness designation as the solution to our environmental ills.

First, it describes nature as a "life-support system" that provides us with "important resources" and "ecosystem services." Human economic activity (or even social and cultural activity, if it is sedentary) constitutes a kind of suicidal wrench we are throwing in the ecological clockworks that sustain us all.

Second, nature is described as a kind of "therapy" that "rejuvenates our spirit" and provides us with "the simple pleasures of human-powered recreation." Of course, nature has not always been seen this way. Early in this continent's Euro-American history, people celebrated the "City Upon a Hill" as a haven from the dark and dangerous wild without. But then sometime during the nineteenth century, those cozy little cities became stinky, poisonous, violent, and noisy industrial monsters, and all of a sudden it was the "wildness of nature" that seemed more like a haven. The important point here is that, for Euro-Americans, the world has always been split into "civilization" and "nature," and that what we now call "environmentalism" was made possible by simply sliding nature from the bad to the good side of the split.

Finally, nature is described as a kind of "citizen" whose value should have nothing to do with its usefulness to human beings, as one of many competing "interests" with all the rights that citizenship grants (but not, you'll notice, any of the responsibilities). "We've done pretty well including humans in our moral contracts," proponents argue; "now it's time to include nature."

All three of these metaphors, while they seem on the surface to be promoting greater respect for nature, rely heavily on the classically Euro-American separation of human communities and the ecosystems in which they are embedded. The first two make it very clear: Nature is a "resource," a pool of stuff "out there" that can help us with our own physical, emotional, and spiritual health. The third, while it disassociates the value of nature from human needs, it also pits human communities and nature against each other in a Hobbesian legal struggle of all against all. Worse yet, it seems to be saying to nature, "We realize that the concept of rights has not occurred to you, or that you have found no use for it, but we have concluded that it is in your best interest, so we are going to force it on you."

In the case of the ONRC's Oregon Wild Campaign, the connection between its particular brand of nature appreciation and the traditional American segregation of civilization and nature is just too blatant to ignore. How else could the ONRC see a solution to the problem of "mankind's alienation from nature" in the bureaucratic designation of wilderness areas? How else could it see the systematic exclusion of human communities from certain landscapes as the only way to foster a sense of respect for those landscapes?

The ONRC is able to come to these conclusions because, frankly, its members live in cities and suburbs, where the suppression of wildness is so basic that even they don't notice how thoroughly it influences their lifestyles. Whether or not they like to think so, they have bracketed off their own lifestyles—relentless wonderlands of technological artifice and corporate insinuation—as the "norm," as the only possible way for humans to live in this day and age. They reluctantly resign themselves to the fact that their own "places" are sort of industrial sacrifice areas, beyond hope. They recognize that the human animal needs a closer connection with nature than urban life provides, so they resolve to make sure there is some wilderness out there "somewhere."

But even for those hapless urbanites sitting at computers on the fourteenth floor, there is wilderness all around them and, for that matter, all through them. Somewhere fourteen floors down is some dirt; and mixed in with that is the residue of the plants and animals and human communities that that dirt used to support. Somewhere under that asphalt you drive home on are biota frozen since the machines rolled over, some seeds waiting for sunshine. And I suppose we all know that one of these days they'll get what they're waiting for.

What I'm trying to say is that it's just plain lazy to think of nature or wilderness as something that exists only somewhere else. Yes, you urban folks have done a much better job mangling native landscapes than we rural folks have, but I guess I don't see that as a good reason to write them off, spending all your time and effort trying to preserve distant landscapes, just so you can go there once in a while and remind yourself how far gone your own home is.

There's that word again: home. That shimmering mist of people and place all drawn together by shared stories. You have it there, in the cities and suburbs, whether you see it or not. Nature, "untrammeled by man," is under your feet, over your head, and bubbling all through your hearts and minds, but for all the reasons we've talked about, you feel the need to find it elsewhere.

I suppose it is possible to "love" a federally designated wilderness area, but as seems clear from the Adopt a Wilderness program, it will be the condescending love of a parent for a child. Or it will be a kind of puppy love that focuses on the "simple pleasures," rather on than a living, breathing grownup love that keeps front and center the fact that, as Gary Snyder put it, "there is no death that is not somebody's food, and no life that is not somebody's death." It's a good bet whoever wrote "Home on the Range" way back when lived this truth every day of his life.

A federally designated wilderness area can never be home to anybody—any human being, anyway. I realize that's the point, but I guess it just feels like we're faking it. In Klamath County, some farmers and ranchers have recently established a watershed council, and one of the ways we describe the work we do is simply "good housekeeping." We do this work not to save the Earth, and not to feed the world, but for the same reason we do our dishes, sweep the floor, mow the lawn, and dust our Bibles. We live here, and we need to keep things in good working order. It's our home, and we tend to it. This landscape of ours is the here and now that tells of generations of shared history, and every time someone says to someone else, "Remember when that happened there?" our local communities settle a little bit more into place.

There's that other word: place. I think it's the same thing as home. And I guess I'm suggesting to members of the ONRC that rather than being "parents of place," consider being married to it; that instead of neglecting your own homes for the sake of occasional casual affairs with some distant landscape, give yourself up to your own landscapes "for better and for worse." Instead of mapping potential wilderness sites, start mapping the landscapes where you live, and instead of interpreting data, start interpreting the stories and histories that connect you to your home and to the people with whom you live and love.

I read an amusing account of the 1993 Land, Air, Water Conference in Eugene. It was included in "A New Challenge: Overcoming Green Sexism," an article written by some ONRC staff members. The authors recount that "while the organizers did include a committee on sexism, the conference was disrupted by a group of Native American drummers protesting the all-white makeup of the televised keynote panel." Now, I don't mean to disrespect the histories of either women or Native Americans, but I couldn't help but be reminded of Monty Python's *Life Of Brian*, in which the Judean People's Front, the People's Front of Judea, and the Front for Judean People argue bitterly over who are the more oppressed. The solution they come up with is, of course, to call another meeting immediately.

My point is that environmentalists seem to spend as much time and energy fighting each other as they do trying to save nature, and I would argue that this is because they are struggling for representation in an abstraction rather than working for a good life on the land. If you let it, the land will take hold of you, and the things that happen there will become stories that will bring you together.

Jim Corbett, a rancher in the Southwest who helped smuggle El Salvadorans into the United States during the 1980s, spoke to this in Sharman Apt Russell's unfortunately titled book *Kill the Cowboy*: "Today, the environmental movement has all sorts of different agendas pulling people in different ways. . . . Just as heedlessly as industrial civilization displaces untamed biotic communities, it is burning its bridges with livelihoods that are rooted in humanity's pre-industrial relations with the land. Keeping our options open by preserving these livelihoods is a matter of ecological wisdom, not just an exercise in nostalgia." Russell herself points out that "a very, very good rancher can out-environmental the environmentalists every time."

Aldo Leopold, whom the ONRC is fond of quoting when promoting wilderness designation, argued that "the first law of intelligent tinkering is to save all the parts." He also claimed that

> Perhaps the most serious obstacle impeding the evolution of a land ethic is the fact that our educational and economic system is headed away from, rather than toward, an intense consciousness of the land. Your true modern is separated from the land by many middlemen and innumerable physical gadgets. . . . Turn him loose for a day on the land, and if the spot does not happen to be a golf links or a "scenic" area, he is bored stiff. If crops could be raised by hydroponics instead of farming, it would suit him very well. Synthetic substitutes for wood, leather, wool and other natural land products suit him better than the originals. In short, land is something he has "outgrown."

We farmers and ranchers would like to keep our land and our people the way we would keep wearing an old shirt, or the way we feed a good horse years after it's quit making money for us. And I wish the ONRC would start seeing us as people it could learn from rather than enemies to be conquered. And if the ONRC must be everywhere all the time, how about working with us instead of against us?

The ONRC also loves to quote the last paragraph of Wallace Stegner's "wilderness letter," in which he calls wilderness "the geography of hope." But I've noticed the ONRC always forgets about the paragraph before that, in which Stegner says he has seen enough range cattle to recognize them as wild animals; and the people who herd them have, in the wilderness context, the dignity of rareness; they belong on the frontier, moreover, and have the look of rightness. The invasion they make on the virgin country is a sort of invasion that is as old as Neanderthal man, and they can, in moderation, even emphasize a man's feeling of belonging to the natural world.

So lighten up on us cowfolk for a while, and see what we can do when we put our minds to it.

REFERENCES

Leopold, Aldo, "A Land Ethic," in Roderick Nash, ed., *American Environmentalism: Readings in Conservation History.* New York: Knopf, 1989.

Snyder, Gary, *The Practice of the Wild.* Berkeley, California: North Point Press, 1990, 184.

What Do We Mean by Consensus?
Some Defining Principles

Matthew J. McKinney

During the past five years or so, the philosophy and strategies of collaboration and consensus building have come to the forefront of discussions on
natural resources and public policy in the American West, largely in the context of community-based forums. This growing body of experiments—community-based conservation, watershed councils, forestry partnerships, study
circles, and other types of community forums—is defined by the voluntary
engagement of people with diverse viewpoints in sustained conversations
over the social, economic, and environmental values of a particular place,
such as a watershed, river basin, ecosystem, or rural community. Today it's
hard to find a community or bioregion in the West that does not have some
type of grassroots, citizen-driven study group, community forum, or citizen
council.

Some people argue that these initiatives reflect a revival of Jeffersonian
democracy in the American West. More than two hundred years since the
Founders framed our representative democracy, Donald Snow, Daniel Kemmis, and others argue that this movement signals a philosophical shift from
Madison's belief in the ability of government agencies to arbitrate among
competing interests to Jefferson's view of the fundamental need for delibera-
tive forums where citizens and officials work together to shape and implement public policy.

Community-based forums may be the logical extension of decentralizing decision making, from the federal government to the state government to the local government to the citizens. These efforts reflect a renewal of citizenship, and complement the efforts of government officials to meaningfully involve citizens in resource decisions.

This diverse body of initiatives is often referred to as part of the "consensus movement." Such a label, however, does not accurately reflect the intent of many collaborative processes, and creates unrealistic expectations that are rarely met. The consequence is a growing sense of skepticism and concern about the appropriate use of collaboration and consensus in natural resource policy.

To inform, invigorate, and improve the effectiveness and satisfaction with these emerging forms of governance, it is helpful to consider the defining characteristics of consensus processes. This reflection, in turn, will help distinguish what is and is not a consensus process, and clarify how these processes fit into the mosaic of ways that citizens and officials come together to solve public problems.

To begin, let me suggest that there are four basic approaches to making public decisions and resolving public disputes. The first approach is to determine who is more powerful. Power in this sense is simply the force of coercion: the ability to make others do something they don't want to do. Natural resource politics in the West and elsewhere is replete with examples of the use of coercion. A great deal of legislation and regulation is characterized as "command-and-control," meaning that it imposes some type of action and associated cost upon others. Other power-based procedures are equally coercive. Civil disobedience, strikes, elections, and citizen initiatives are all examples of ways in which we try to solve problems, keep bad things from happening, or promote policy objectives through the use of power.

A second approach to making public decisions is to determine who is right. In this context, we usually rely upon some independent, legitimate, and fair standard to determine rights or rightness. Some rights are formalized into law or contract. Others are generally accepted standards of behavior, such as reciprocity, precedent, equality, or seniority. Rights, however, are rarely as clear as they may seem.

The underlying problem of instream flow protection, which I refer to later, is a good example of how a rights-based procedure works. In this case, farmers and ranchers hold senior water rights that allow them to divert water out of a stream, which may hurt the fishery resource. While conservationists and others may complain about dewatering, they have few if any rights to keep water in the stream. Decisions on the legal allocation of water rights often involve a third party, such as a court or an administrative agency, with the power to hand down a binding decision.

Rights procedures, much like power-based procedures, are adversarial in nature and tend to result in winners, losers, and impaired relationships among the parties. In both cases decisions are made, but the underlying problems are rarely resolved and the losers do not "go away."

A third approach is to reconcile interests. This is where collaboration and consensus building come in. The essence of collaborative decision making is to reconcile the interests of affected parties, sometimes referred to as stakeholders. Interests are needs, desires, concerns, and fears, the intangible items that underlie people's positions or the items they want. Reconciling interests is not easy. It involves probing and examining concerns, devising creative solutions, and making trade-offs to accommodate competing interests. The most common procedure for doing this is negotiation, the act of back-and-forth communication intended to reach agreement.

Contrary to some people's impression, collaboration is not something you do with the enemy to betray your friends. Rather, it refers to a process whereby a group of people work together to achieve a common purpose and share resources. Collaborative processes may be more or less inclusive, depending on the intent of the participants, and may or may not rely on consensus as a way of making decisions. Collaborative processes—such as watershed councils, community forums, and study groups—are increasingly convened by citizens. Government officials are also convening more and more advisory councils, task forces, and study groups that incorporate many of the defining characteristics of collaborative processes.

Consensus building is perhaps best understood as a particular form of collaborative problem solving, meaning that consensus shares many of the same traits as collaborative processes. However, whereas collaborative processes may or may not be designed to reach agreement or resolve a dispute, consensus processes are designed to seek agreement among all the people interested in or affected by a particular issue or situation. Consensus processes seek unanimous agreement, in contrast to processes where a decision is reached through voting (a power-based procedure where the majority rules) or is made unilaterally by a government agency or judicial body (typically a rights-based procedure). Consensus processes are further defined by an emerging set of defining principles or key ingredients.

Before we examine these defining principles, it is helpful to put into perspective a fourth approach to making public decisions, one in which decision makers consult the public. Public involvement, public participation, or citizen involvement often starts by informing and educating citizens about a proposed decision or action through press releases, videos, public service announcements, and public meetings. These tools tend to be unidirectional: they tell citizens what is happening but do not provide an opportunity for citizens to respond. They rarely, if ever, allow citizens to provide information or contribute to the education of all affected parties, including government agencies.

The input and advice of citizens may be necessary to develop effective public policy, but they are rarely sufficient to build agreement among diverse interests. Because of the diversity of viewpoints expressed during public involvement processes, government officials typically receive competing, conflicting ideas on what to do. It is then up to them to make the necessary trade-offs among competing viewpoints and to render a decision. As we all know, citizens tend to challenge both the process and the outcomes of most public involvement processes.

Each one of these approaches to making public decisions and resolving public disputes is appropriate, depending on the objectives of the stakeholders, particularly government officials. The challenge for people interested in sustaining the land, water, and communities of the West is to appreciate the differences and consequences of these different approaches, and to select the most appropriate process for a given situation. As one colleague suggests, we need to "match the forum to the fuss."

Key Ingredients for Consensus

While no clear model exists for building consensus, there seem to be at least seven key ingredients for success. These ingredients not only provide a road map for how to design and manage a successful consensus process, but also serve as defining principles. Think of them as benchmarks for determining what is, and is not, a consensus process and how consensus processes differ from other forms of public consultation and dispute resolution.

At the outset, it is important to acknowledge the difference between a consensus process and a consensus outcome. It is possible, and quite common, to convene a group of people with diverse viewpoints to seek consensus and not arrive at a consensus outcome. In other words, consensus processes, as defined here, do not always result in consensus outcomes. That's okay, as long as we understand and acknowledge the difference. Even when they do not result in consensus outcomes, consensus processes often result in many benefits, such as improved working relationships, a better understanding of the issues and different people's viewpoints, and the pros and cons or consequences of different options on how to proceed. A common problem emerges when some participants characterize the product of a consensus process as a consensus outcome, when in fact the product was less than unanimous agreement.

Building on that distinction, here are the seven ingredients for a successful consensus process:

1. Agree on the Purpose. The first thing to realize about consensus processes is that people need a compelling reason to participate—they must agree on the need to change the status quo. They also need to agree that a consensus process is the best approach in that situation. This suggests that consensus processes are not a panacea to public decision making, and that under certain circumstances other public involvement or dispute resolution methods may be more appropriate. When all sides come together with a common purpose, however, the results can be dramatic.

In 1994, after nearly thirty years of conflict and debate on how to maintain and enhance stream flows in Montana for fisheries, individuals and organizations representing farming, ranching, recreation, and wildlife interests came together to seek agreement on how this issue might be resolved. The participants seemed tired of fighting; all of them realized that they might get something from a consensus process that they had not been able to get through legislative, administrative, or judicial processes. After nearly a year of negotiations, with the assistance of the Montana Consensus Council, the participants agreed on a bill that easily passed the 1995 legislature—to the amazement of nearly everyone.

2. Ensure the Process Is Inclusive. All of the individuals and organizations affected by or concerned with the issues being addressed should be involved in the consensus process. This "constituency for change" should include people affected by any agreement that may be reached, those needed to successfully implement an agreement, and those who could undermine one if not included in the process.

Before convening the instream flow negotiation, I spent several months visiting with people interested in the question of instream flows. The objectives of these conversations were to clarify the interests of the different parties, determine if they wanted to participate in a consensus process, and identify other stakeholders. Interestingly, the participants quickly agreed that while both the legislature and state agencies would more than likely be needed to implement any agreement, they were not welcome at the table. The participants feared that having legislators and agency officials involved at the outset would unnecessarily politicize the conversation and create artificial administrative barriers to innovative solutions.

In several other projects I am involved with, the core participants—citizens and advocacy organizations—are likewise choosing not to include politicians and agency personnel. I'm not sure what to make of this trend, except to note that it cuts against a pure theory of absolute inclusiveness. In certain situations, there may be legitimate reasons for not inviting selected stakeholders to participate directly in the consensus-building process.

3. Allow Participants to Design the Process. One of the key defining characteristics of a consensus process is that the participants design and manage it. This principle, more than any other, distinguishes a genuine consensus process from other types of public decision making. Participants must be free to design a process that fits the specific needs of their situation. They should select their own representatives, define the issues in their own terms, and develop appropriate ground rules to govern the conversation.

Legislatures, state and federal agencies, and local government bodies cannot and should not force people with diverse viewpoints to participate in consensus processes. Such mandates are contrary to the basic idea that consensus processes are "voluntary." Government authorities may suggest the use of consensus processes to address particular issues, but if the stakeholders are not interested or are not given the freedom to design the process, it is not, in the pure sense, a consensus process. In this respect, most of the advisory committees and task forces appointed by government officials are not genuine consensus processes. This is not to say that such processes are not useful: the distinction is important to help clarify the different expectations and outcomes of different processes. A genuine consensus process, where all the stakeholders—including government officials—select their own representatives and jointly establish sideboards, results in a much different set of expectations and outcomes than a public consultation process where a limited set of stakeholders—either an agency or group of citizens—does its very best to handpick credible, legitimate people that represent the diversity of viewpoints and unilaterally impose sideboards on the conversation. In the latter case, it is quite common to observe groups where the stakeholders have little or no ownership of the process, and it is not clear whether the participants are representing themselves or some larger constituency. As a practical matter, it also begs the question of whether a consensus outcome is an agreement among the people at the table or among some larger group of constituents. Finally, the top-down, handpicked approach often results in situations in which people who have not been chosen, but clearly have an interest in the issue, demand an explanation or seek to disrupt the process.

Another key step in designing a forum is to agree on what consensus will mean, and who qualifies as a "consensus builder." In many of the groups we work with, the participants adopt the following definition of consensus:

> Each participant is committed to seeking consensus (or agreement). Consensus is reached when the participants agree on a package of provisions that address the range of issues being discussed. The participants may not agree with all aspects of an agreement; but they do not disagree enough to warrant their opposition to the overall package. Each participant or stakeholder group:

1. Has the ability to disagree with any proposal, but assumes a responsibility for providing a constructive alternative that seeks to accommodate the interests of all participants;
2. Is committed to implementing agreements that are reached; and
3. Will maintain their values and interests.

For purposes of this dialogue, agreement is defined as agreement among the participants or stakeholder groups.

This decision rule implies that consensus processes do not require participants to compromise their values and interests, and that each participant has a right to veto any proposal made during the conversation, but assumes a responsibility to provide an alternative proposal that seeks to accommodate the interests of all participants.

4. Encourage Joint Fact Finding. By gathering, analyzing, and interpreting data together, participants build a common understanding of the situation. Everyone is encouraged to share his or her unique interpretation of the information, and all should have equal access to it. Building a common understanding seems to be a prerequisite to building agreement.

5. Insist on Accountability. To ensure success, the participants in a consensus process must respect the forum they have jointly created and assume a sense of obligation to maintain its integrity. At the same time, it is absolutely essential that participants report to their constituencies in a timely and consistent manner, and relay the concerns of their constituents to the other members of the forum. The goal of a consensus process is not to reach agreement among the negotiators per se, but among the constituents they represent. This principle also suggests the need to keep the public informed of major accomplishments.

6. Develop an Action Plan. Ultimately, the participants must commit to a plan of action. This includes identifying roles and responsibilities to implement the agreement, designing a strategy to monitor and evaluate the agreement, and planning to reconvene the group, if needed, under specified circumstances.

As the agreement on instream flows began to take shape, the participants worked to ensure that the agreement would be accepted by the legislature. They identified who should lobby which legislator, the order of testimony during committee hearings, and even the content of the testimony. They also agreed to form a working group to monitor implementation of the new law.

The working group, with the assistance of the Montana Consensus Council, has produced a brochure explaining the new instream flow law. Trout Unlimited, one of the organizations that negotiated the agreement, is pursuing several instream leases to maintain fisheries in Montana and has already learned a great deal about the process of changing water rights from historical uses to instream flows. And if a problem emerges, we have a built-in mechanism to reconvene in order to find a solution.

7. Find the Collaborative Leader. Another important ingredient to the success of consensus processes is the need to find a credible, legitimate leader to coordinate and lead the process. A collaborative leader contributes a range of important services.

First, a collaborative leader can help stakeholders assess the situation, design an appropriate forum, coordinate meetings, document agreements, and support the action plan. Second, with no stake in the outcome, the facilitator can help legitimize the process, improve communication among participants, and ensure enforcement of ground rules designed to equalize power. Finally, a collaborative leader should also contribute to the integrity of the process by identifying unrepresented interests that emerge during the building or implementation of any agreement, and suggest ways to incorporate such interests.

Why Use Consensus?

As consensus processes continue to emerge and are shaped by those involved in them, critics of the effort have leapt to the fore. Some people claim that consensus processes simply do not work and are inherently biased. The Oregon Natural Resources Council (ONRC), for example, maintains that such processes "include interests which benefit from the status quo, and therefore have little or no desire to change." The ONRC goes on: "Our participation is sought most to co-opt us into a process where it is assumed we would lay down our legal and other weapons to talk in good faith."

The Montana Environmental Information Center claims that consensus processes are "fundamentally undemocratic, for the process grants tyrannical power over the majority to the minority." In what must be one of the broadest and most scathing indictments of consensus-based approaches, former British Prime Minister Margaret Thatcher has said, "Consensus is the process of abandoning all beliefs, principles, values and policies in search of something in which no one believes, but to which no one objects."

As mentioned earlier, consensus is not a panacea. A consensus process is appropriate when all of the stakeholders believe that they are likely to get something through consensus that they are not likely to obtain from any other arena. For those who believe that their interests will be better met through the courts, the legislature, or some other avenue, I suggest that they consider four questions:

1. What are the likely costs—the time, money, and emotional energy—in relying on one process relative to another?
2. How satisfied are you likely to be with the outcome of a particular process?
3. How will the use of one process over another impact your relationships with other affected parties? Do those relationships matter in the long run?
4. What is the chance that the issue will actually be resolved through one process or another? What is the stability of the likely outcome?

When properly used in appropriate situations, I propose that consensus is generally less costly than determining who is right, which in turn is less costly than determining who is more powerful. Consensus is also more effective than traditional public involvement processes when the goal is to seek agreement on public policy. In my experience, consensus approaches tend to produce wiser, more satisfactory outcomes; reduce costs; build relationships; minimize recurring disputes; and result in effective public policy. Given the growing interest in the role of collaboration and consensus in western resource policy, it seems to me that we need to further clarify, discuss, and seek a common understanding of what is and is not a consensus process, and to distinguish these processes from other approaches to public decision making and community-based conservation. The principles presented here provide a place to begin this conversation. Additional defining principles may emerge from our ongoing experiences.

DEFINING THE TERRITORY:
THE CHANGING FACE OF
THE AMERICAN WEST

Collaborative conservation did not emerge in a vacuum. Rather, a variety of demographic and economic changes over the past couple of decades have changed the face of the West. The land of the range-riding cowboy (or at least of the cowboy myth) is now the domain of the Web-surfing modem cowboy, living in his 5,000-square-foot starter castle built to capture prime wilderness views. Americans and foreign visitors with growing disposable incomes have flooded into the western public lands to mountain bike, ski, hike, ride all-terrain vehicles and snowmobiles, and otherwise engage in what is increasingly seen as industrial tourism. This section's lead essay, by William Riebsame, outlines the major changes taking place and speculates about the impact of newcomers' attitudes on resource management and conservation efforts.

At the same time, the policy arena in which decisions are made about the West's substantial public lands has undergone equally dramatic changes. A combination of global economic forces, automation, and national environmental challenges resulted in substantial changes in public resource management—displacing workers and leaving some communities wondering about their futures. And while some economists assure local leaders that the rapidly growing service economy is far outpacing losses in the resource extraction fields, not everyone is sanguine about the changes. The paired essays by Ray Rasker and Lynn Jungwirth well illustrate the current dialogue about the region's economic future, taking into account the impacts these changes have on the individual and his or her community.

Peter Decker's essay assumes the economic and demographic changes described previously, but goes on to argue the existence of enduring regional traits and values—a Code of the West—including the traditional habit of "neighboring," which provides the foundation for collaborative conservation as well as other rural survival strategies.

In light of the many changes underway in the region, this section concludes with a speculation into the nature of "community" by Carl Moore. Like Decker, Moore argues that we make sense of our own lives through meaningful interactions with others, and he suggests a definition of community to help us understand the various ways people in this region have worked together despite significant differences. As reflected in the title of the *Chronicle of Community*, we believe that the notion of "community" is the center of the growing collaborative conservation movement in the West—not a utopian ideal of "everyone getting along," but a recognition that people who work to build relationships with one another are capable of pursuing common goals despite diverse individual values.

Geographies of the New West

William E. Riebsame

Another great wave of development washed over the American West in the 1990s and, at this writing in late 1999, probably has yet to crest. The region has experienced booms before, and each rearranged its geography. We might even mark the initial development boom at some 13,000 to 15,000 years ago when the first people made it to North America, and in the course of a millennium or two managed to occupy the whole region, at least seasonally. Certainly, the Spanish conquest from the south in the 1500s constituted another regional "boom," as did the gold rushes, the Mormon settlement of the Great Basin, and the cattle, irrigation, military, and energy booms of the nineteenth and twentieth centuries.

Booms, of course, imply busts. Falling global oil prices deflated the last regional boom in the early 1980s, and parts of the West experienced net out-migration—more people leaving than moving in. The office towers of Denver and the subdivisions of Rock Springs went from chock-a-block full to a third empty in just a couple of years. Yet no bust is total; each cycle prefigures the region's geography for the next development wave. The cheapest office space in the country attracted new businesses to western cities in the early 1990s, and the national economic surge, partly supported by the low energy prices that busted the West in the 1980s, nurtured the region's 1990s high-tech and quality-of-life florescence. We make and remake the region, in a pattern more cumulative than boom-and-bust.

The New Gold Rush

The latest wave of western development is, at the turn of the millennium, setting records for the regional immigration it has occasioned, for the new homes and businesses built, and for the jobs, income, investment, and wealth it has brought to the region. But the latest boom cannot be analyzed like previous ones—perhaps should not even be called a boom: no single commodity, or even suite of related commodities (like uranium and strategic metals) drives it; it is not focused on a particular resource, not anchored to the subterranean geography of oil fields or coal seams. Nor is it occasioned by a particular innovation or singular economic compulsion. Instead, the current boom looks to be a totally different regional experience, something approaching a maturing and diverse development pattern that appears to combine powerful elements of technology, investment, entrepreneurialism, and inherent regional qualities. This boom hints at a sustained regional florescence, a new geography so attractive and so well suited to the structure of the postindustrial economy that it could someday dominate the economies of other U.S. regions.

Time magazine heralded the boom in its September 6, 1993, cover story, "Boom Time in the Rockies," which described families escaping coastal and Midwestern urban blight to take high-tech jobs in the West—and watching sunsets behind snowcapped mountains. The story cites Denver's new mega-airport and high-tech office parks, Salt Lake City's booming economy and award of the 2002 Winter Olympics, Montana's growing service economy, and Jackson Hole's housing shortage as indicators of the region's vitality, comparing these trends to the mid-1980s energy and mining bust that all but emptied Denver's high-rises and briefly made real estate affordable in Crested Butte, Bozeman, Ridgeway, and other mountain towns now too expensive even for upper-middle-class home buyers.

Seven years later the boom has not slowed. Phoenix and Las Vegas vied to rank as the fastest-growing American city in the 1990s—Phoenix won, growing 21 percent (200,000 residents) during the period 1990–98. New home sale and price records were set in Denver (prices climbed 17 percent from 1998 to 1999), thousands of new residential units sold even before they were built in the booming ski resorts, and ranchland pretty much everywhere in the West fetched twice to ten times its agricultural value.

Only a few places lag behind. Wyoming struggles with slow growth, and overattention to the oil rig count as a measure of economic success; its population grew only 6 percent during 1990–98, slowest of the eleven western states and more akin to the stagnant pattern of Great Plains states like South Dakota. And a few western towns still suffer from loss of the extractive industries—the last mine in Leadville, Colorado, closed in July 1999, the same month that copper operations slumped across Arizona and New Mexico, and the second-largest gold and silver mines in Idaho had closed the previous fall.

Still, the main story in the West is rapid development, and the current expansion has some qualities of the previous booms: it attracts immigrants from all over the United States and internationally; it creates tensions between the newcomers and longtime residents; and it raises many of the enduring debates about access to land and resources. But its geography is different, more pervasive. The number of people living in the region matters less than where and how they live, and this latest western boom is remarkable for deploying more people and development in all of the region's main landscape settings: core downtowns, suburbs, and out in the vast open spaces, near federal lands and wilderness areas.

New Geographies

Five distinct land-use patterns beckon our notice in the modern West: city sprawls, ex-urban sphere, gentrifying range, federal commons, and resort zones. The city sprawls cannot but first attract our attention; the region hosts the fastest-growing, most sprawling cities in the United States. Los Angeles, Phoenix, Las Vegas, and Utah's booming Wasatch Front glare out from nighttime satellite images against a black background of open spaces. Even smaller towns like Bozeman, Boise, and Bend are sprawling in the New West. Western observers point out that the West is "urbanized": most westerners live in urban areas. But what does this mean geographically? First, most westerners are not urbanites, they are suburbanites, and the suburbs, not the central cities, saw most of the region's growth since World War II, and are now gaining most of the new jobs and businesses. During 1990–98, the Phoenix suburb Chandler grew 78 percent (70,000 residents), and Henderson, outside Las Vegas, more than doubled while Mesquite, Nevada, grew by 441 percent from 1,821 folks into a fair-sized suburban town of 10,125. But while such suburbs will account for the majority of the 25 million more people expected to live in the eleven western states by 2025, there is some action in the old core cities. Several "down towns" have reversed the drain to the suburbs. Denver proper—a geographically small city-county surrounded by suburban towns—lost 25,000 residents in the 1980s, but gained 29,000 during 1990–98. Core-city residences of all sorts, from Victorian cottages to lofts, now sell at a premium.

Beyond the city sprawl lies an emerging ex-urban sphere, or exurb, a newer land-use pattern in which the rural and suburban mix it up, where areas well removed from the city feel the influence of a dispersed suburban-like development tied to growing mobility of businesses and incomes. John Cromartie, a geographer with the U.S. Department of Agriculture, has shown that nonmetropolitan counties adjacent to metro areas are the fastest-growing places in the United States (think Park County, Colorado, or Summit County, Utah, both relatively short rural commutes from large cities). The exurbs also gained jobs faster (44.5 percent) than metro (26.6 percent) or more rural counties (32.5 percent) during 1985–95 in the West, according to geographer William Byers. In a sense, these adjacent, nonmetro places are now really all part of the "urban" geography of the West—their populations rely on the nearby cities if not for daily jobs, then at least for their urban services: hub airports, entertainment, venture capital, banking, and the like.

Out past the exurbs lies the gentrifying range, a deeply rural yet subdividing landscape now searched out by hobby ranchers and other New West homesteaders who appreciate the wide-open spaces. Once upon a time in rural studies, population increases meant only one thing: a growing farm, ranch, timber, or mining economy. Now we find those economies dying, yet all but a handful of rural counties in the West growing; indeed, on average they're growing faster than metro counties. Cromartie, the USDA analyst, found that nonmetro counties in the West grew 15.5 percent during 1990–97, while metro counties grew an average of 12 percent. This gentrifying range is tough to figure out. We really don't know how much ranchland is now in nonranching ownership, much less can we readily distinguish between working and hobby ranches. Some of the new houses that dot the range are homes to the ranchers; some bespeak a local from town who has made it good enough to buy acreage. Others are second, third, and fourth homes for the itinerant wealthy, some are retirement spots, and many are the office/residences of a newly mobile professional class that appreciates the West's open landscapes. Even these truly rural places gained jobs faster (32.5 percent) than metro areas (26.6 percent) in 1985–95, according to geographer Byers, outpacing the national average job growth of 23.4 percent.

The gentrifying range abuts the vast federal lands that still mark the West as a unique American geography. This federal commons offers an increasing contrast, and a tension zone, with the rapidly developing private lands. The public lands are so well studied and written about that their contemporary development does not bear repeating here. But the gentrifying range and federal commons are perforated by another impressive development landscape, the resort zones like Aspen, Jackson, and Sedona—where industrial tourism and landscape amenities inflate land values and bring extravagant new commercial and residential development to mountain and desert landscapes.

Driving and Enabling the New Geographies

A coalescence of driving and enabling forces shape these development geographies, guiding capital to New West real estate. The booming national economy, international immigration, and high birth rates impel Western development. A burgeoning love of western landscapes, and long-standing American preferences for residential space that drove the postwar suburban boom further, drive the New West's sprawl. The region suits America's postindustrial sensibilities. The West is the nation's green-field suburb, where development is easy, relatively cheap, and quite profitable. Income and corporate mobility, employment flexibility, retirement, and communications, transportation, and construction improvements all enable the new homesteading, especially its exurban and rural deployment.

The coalescence of forces for development, which often disregards community desires, has been called the "growth machine," and the machine is humming along nicely in the West. The well-known cogs in the machine— property rights, investment strategies, profit margins, government subsidy, growth-oriented land-use planning and zoning, and the notion that any community not growing is dying—are now increasingly driven by an additional fuel: landscape preferences. Economists Thomas Power and Ray Rasker tell us that the West's new economy is driven by people seeking a higher quality of life. They go where they want to live and bring development with them. A large corporate example of this sprawls on the highway between Boulder and Denver. Sun Microsystems is adding some 3,000 employees to its new campus in the Interlocken office park; and a two-year-old start-up will house 1,500–2,000 employees there. Level 3's CEO reported that he chose the Colorado Front Range location because a survey of 500 college seniors and current high-tech employees told him where and how his potential workforce wanted to live. He told the *Denver Post* that the Denver area beat out Boston, San Francisco, and the other usual locations. Our own preferences power the growth machine, and we guide corporate strategy and transform regions.

Limits on Development?

The economics, demographics, and lifestyle preferences of a mobile, wealthy population eager to become western, would seem to ensure much more regional development. But are there limits on western growth, some absolute regional carrying capacity? Somewhere buried in the psyche of most westerners is a sense that urban growth—or regional population growth overall— will eventually be limited by some geographical factor, with water, or rather, lack of water, taking pride of place on this list of perceived barriers. Every schoolkid knows that the West is arid (or at least semiarid) and that cities like Phoenix and Salt Lake face potentially crippling water shortages that will ultimately halt, and maybe reverse, their growth.

Every schoolkid is wrong. Plenty of water is already developed for decades of future urban growth in the West, far too much water for climate to act as a brake on development. The notion that the West could not develop as intensively as the East got started with Kit Carson, John Fremont, and other early Europeans who labeled the Great Plains and Rocky Mountain area the "Great

American Desert." John Wesley Powell made explicit the notion that aridity would and should limit Western development, and offered planning guidelines limiting population and agricultural development to suit regional water supplies. He even proposed laying out governmental jurisdictions along watershed boundaries, implying that water was the most important factor in social organization. The actual history of urban growth in the West, however, has proved to be much more traditional: what matters is good old-fashioned urban boosterism, jobs, investment, tax base, and ability to lure government subsidies and the next industry as well as to annex the land on which the next office parks, subdivisions, and malls will be built. Water hardly makes an appearance as a shaping, much less limiting, force.

When colleagues and I at the University of Colorado put together the water chapter in the *Atlas of the New West*, we made a simple calculation showing this. We noted that 80–90 percent of western water was used in agriculture, but that we could find a few places, like part of the Colorado Front Range served by the Colorado–Big Thompson water project, where almost half of the agricultural water had shifted to municipal ownership. This is far ahead of most places (only 2 percent of Idaho's water use is municipal and industrial), but points out the potential. If half the agricultural water use in the West shifted to urban uses, the region could support, at current rates of use, a fivefold increase in urban population, and this without extracting a single new drop of water from streams and aquifers. The West is too wet.

Now, many other things might cap western development: air pollution, species loss, traffic, even cultural tensions. And the transfer of water from agriculture to urban uses, even with willing sellers, is not painless. But the analyst driven by history, rather than by prediction and prescription, must allow that Los Angeles, Phoenix, Atlanta, and other booming American cities exist for enduring reasons, what you could call geographical laws of urbanization. I see no reason to expect that these principles are now in abeyance in the interior West. Finally, if locational preferences are driving development, as they seem increasingly to be, then a slowdown comes only when those preferences change, or when the region no longer meets them. Both of these watersheds are a long way off in the West according to my crystal ball.

Conservation in the New West Landscapes

From Denver to Tahoe, from Tucson to Missoula, the West is developing rapidly, its farms and ranches replaced by subdivisions, its small towns and big cities growing fast, and even its recent immigrants expressing, within just a few years of their arrival, a sense of loss as the places around them change. Much of this change is the papering of western open landscapes with development. It may seem curious that the region with the most open space in the lower forty-eight states is engaged in a debate about land development, but the logic is clear. Much of the region's attraction is its open landscapes, and every undeveloped parcel, public or private, is already somebody's viewshed, someone's open space. Thus, even as growing populations fill the West's landscapes, they will demand open-land protection.

How will the collaborative conservation movement deal with the West's developing geographies? I am better at describing than predicting, but great potential exists, as this book attests, to nurture the West's nascent collaborative conservation movement into an effective force for land conservation. But some of the same forces driving the West's development boom, and many of its social outcomes, pose large barriers to conservation. First, of course, stands our own ambivalent relationship to natural spaces. We want them protected, but we also want to use them, to see them out of our windows, so we settle close to them.

The complex geographies described above also simply make land conservation, collaborative or not, more complicated. A couple of students and I once mapped the changing land ownership of a valley in the Colorado Rockies. The Forest Service owned most of the high country, and fewer than a dozen ranches held the valley floor through the 1960s. By the 1990s, a mile of private/public boundary that once brought two land managers (a rancher and the Forest Service) together, now had two dozen landowners on the private side. Like the rancher, the new neighbors have concerns over wildfire, roads, hiking trails, and cattle grazing, but now dozens of owners' views, not just one, must be incorporated into land planning.

The new homesteaders in the suburban, exurban, and rural West may generally support land and biodiversity protection, but strings come attached to these attitudes. Many New Westerners are politically and economically conservative; many are tax rebels. Most are wary of regulation and outright opposed, in principle, to growth limits. And surely the growing number of absentee land- and home-owners in the West, especially in the resort zones, make community conservation awkward at best.

This book rests on an important assertion: that local, collaborative conservation can work better in the West than so-called command-and-control, centralized conservation. But can community-scale collaboration rein in the American growth machine, now running in high gear in the West? Only with difficulty. Our individual preferences coalesce with national and global economic rationales to drive development in ways that our community preferences abhor. The driving and enabling forces creating new development landscapes in the West operate at multiple scales, and entwine both self-interest and interest group dynamics. The challenge to community-scale, collaborative conservation, then, is to weaken the hold of not only national government but also multiscaled development forces, while simultaneously wedding individual preferences to community desires. A tall order, indeed.

Your Next Job Will Be in Services. Should You Be Worried?

Ray Rasker

I am one of a group of geographers, economists, and others who work with rural communities in the West, helping them understand the makeup of the economy. Among this disparate group of people there is no clear consensus, and if you were to get us in a room together, there would be a lively debate. We clearly don't agree on everything. However, on one theme we all agree: Your next job will be in services. Is this cause for concern?

Those who study the West know that the culture of the region, people's identity, and sense of place are influenced by an economic history that views the land as the primary source of wealth. Translated to bumper stickers: "If It Can't be Grown, It's Gotta Be Mined," or "True Wealth Comes from the Ground." This pervasive belief weakens our ability to compete in a global economy and it impoverishes our communities.

Lester Thurow, a native of Montana and a professor at MIT, offers a different way to think about the economy. In his book *Head to Head*, he points out that the seven key industries of the next few decades are all "brainpower" industries: microelectronics, biotechnology, the new materials industries, civilian aviation, telecommunications, robots and machine tools, and computers and software. Most important, these industries are all "footloose"—they can locate anywhere in the world. And, according to Thurow, "Where they locate depends upon who can organize the brainpower to capture them. In the century ahead the comparative advantage will be man-made."

As any observer of the so-called New West can see, communities like Bozeman and Missoula, Montana, Telluride and Boulder, Colorado, and Jackson, Wyoming, are now full of these footloose entrepreneurs. Roy Romer, the former governor of Colorado, phrased it succinctly: "America's brightest people are attracted by America's most beautiful places."

Check Your Myths against These Realities

The same mantra is repeated all over the West: "If only we could log more trees/open a new mine/get wheat and cattle prices back up, then we'd be okay." For example, Montana Governor Marc Racicot recently attributed the decline in the state's per capita income to "the move away from jobs based on the extraction of our natural resources and toward those in the service and trade sectors." On the surface, it sounds as if we should increase resource production and reduce our dependence on services. It appears that perhaps we should go back to an economy that came of age a century ago—but which we have been leaving behind for the last two decades.

The facts, however, are not encouraging for a return to an old economy. Since 1970 the state of Montana has added more than 150,000 new jobs, and not one of the net new jobs has been in mining, oil and gas, farming, ranching, or the wood products industry. Almost all of the growth has been in services, and what hasn't been in services has been in jobs related to servicing the needs of a growing number of retirees and people who have moved to the state with savings, investments, and private pensions. In short, we have moved from a resource-based economy to a service economy.

Does this mean we no longer make things? Does this mean we no longer create "true wealth"? Does a shift from a "goods-producing" to a "service" economy mean that all we have available to us are dead-end jobs in services? Are we, as it was recently called in *High Country News,* creating a "servant" economy?

The answer depends to some extent on how we define services. This is where some experience with statistics, spreadsheets, and computer databases comes in handy. I like to start with the key point of confusion: The U.S. Department of Commerce organizes its statistical data on employment and income trends into broad categories such as Retail Trade, Construction, Government, Mining, and so on. "Services" is one of those broad, catchall categories. If researchers are not careful, it is easy to misinterpret the numbers.

The Services category includes hotel maids, barbers, and hairstylists—the workers we'd normally think of as service workers—but it also includes engineers, architects, and physicians. Or, as *The Economist* magazine described it: "Policemen and prostitutes, bankers and butchers are lumped together in the service sector." According to Lester Thurow: "Services is simply too heterogeneous to be an interesting category . . . [t]he real issue is not the growth in services but whether the economy is making a successful transition from low-wage low-skill industries . . . to high-wage high-skill industries."

Nationwide, since 1970, the Goods-producing category—primary, manufacturing, and construction industries—has remained stable at around 27 million jobs. Over the same time period, employment in services has risen from 55 million to more than 100 million jobs. In other words, all new jobs created in the United States in the last three decades have been in services.

Does this mean that we are in a transition to a low-wage economy? Some wage comparisons are useful. The average wage of service workers (throughout the Services category, including high- and low-paying jobs) is only 3.9 percent below the average for goods-producing workers, and even that differential is shrinking. In 1979 the median service job paid $82 per week less than the median goods-producing jobs; by 1992 the difference was $19 per week (both of these figures are in 1992 dollars). And when goods-producing versus service jobs are compared in more detail, many of the highest-paying opportunities are found in the service sectors, particularly in transportation and public services (median $605 per week), public administration (median $579 per week), and finance, insurance, and real estate (median $498 per week). By comparison, median goods-producing jobs pay $500 per week.

Perhaps a better approach to the question of services is to do what University of Washington geographer Bill Beyers advocates. He divides services into categories such as producer, consumer, social, and government services. Consumer services generally are low-paying jobs—in amusement and recreation, hotels and lodging, repair shops, motion pictures, household, and personal services. Social services include education and health care. Government services include state and local government, military, and federal employees. Producer services are part of goods production, and include some of the higher-paying sectors—finance, insurance, real estate, legal and business services, membership organizations, engineering, and membership services.

Applying these categories to Montana, more money is earned by people working in producer services ($1.2 billion in 1995) than in all of agriculture (farming and ranching), mining (including oil and gas), and the wood products sector combined ($1.1 billion). The same trend can be seen throughout the intermountain West, where the single fastest source of growth in labor income is in producer services.

Looking Ahead and Adding Value

In light of these economic realities, the important questions for western governors ought to be (1) Are we preparing our workforce to be competitive in a global economy, where knowledge and information are more important than raw materials? (2) Are we preserving the quality of life in our communities to encourage the growth of these industries?

We will realize our fears of losing meaningful jobs if we fail to focus on the training and education necessary to capture the knowledge components of production—design, software development, research, communications, finance, architecture, engineering, and legal work on patents and copyrights. To be blunt, without a concerted effort to provide practical training and education, the higher-paying service jobs—in software development, graphic design, biotech research, and financial specialties—will go to newcomers from the city while locals serve them *lattes* and pump their gas.

Some people become defensive when they learn that we have become a service economy, as if it implies that we no longer need our traditional resource industries. Lester Thurow was recently asked about this when he was the keynote speaker at a convention of manufacturers in Montana. His reply: "Natural resources still matter, but only in the sense of adding value." Given the way our government keeps employment statistics, the value-added component will more often than not be recorded as services, and a diligent researcher will discover that the value-added component can be found in the producer services.

Robert Reich illustrates this point in his book *The Work of Nations*. Imagine, he writes, that it were true that the wealth of an economy depended on the volume of resource materials extracted from the ground. If so, then a bicycle would be valued by the pound, but in fact the opposite is true: lighter bicycles are more expensive. Most of the value—far above the value of the steel, aluminum, plastic, and rubber—is in the engineering, design, testing, styling, marketing, shipping, advertising, and finance. Even producing the tubes for the bicycle's frame involves sophistication, employing metallurgists and structural engineers. You can't make a bicycle without mining, but you can't add value to it without the knowledge industries, or what often gets counted as services. Clearly, both are important.

Which brings us to the question that vexes researchers and community development practitioners: Who wins and who loses when we shift toward a service or knowledge-based economy? Are the better-paying jobs going to those from the city, with experience, education, and connections? So far the evidence is only anecdotal and the answer is mixed. There are good examples of ranchers and rural residents using the new economy and technologies to their advantage, but I know of no rigorous study examining the broader question. Like many, I hope that rural westerners are prepared for more than dead-end jobs as coffee servers and ski-lift operators.

In the Meantime, Growth Happens

Regardless of the unanswered questions about our economy, we should not tolerate hearing our governors, members of Congress, and county commissioners tell us that if only we could return to an economy of the past, everything will be okay. That approach puts blinders on and impoverishes our communities. While waiting for the new mine that was promised, or for the lumber mill to reopen, and while we busy ourselves fighting over mine pollu-tion and clear-cuts, a very different economy is rushing along behind our backs, moving quickly and out of our control. Ignorant of the real economy, and unarmed to steer the growth in a way that benefits us, we lose our ranches and open spaces to sprawl. The roads get congested, and now people seem to be in a hurry all the time (we actually talk about road rage in Bozeman now). The schools get crowded, and city budgets can't keep up with the demand for more snow plowing, better hospitals, more parks, and schools.

In addition to urban refugees moving in, we also see the influx of retirees. They come to enjoy the view and take part in the community. They create new demands for services that, in turn, create new business opportunities. But how will this older population vote? Maybe we won't pass that school bond this year. And who will benefit from this form of growth? Over a third of the personal income in the intermountain West is from nonlabor sources, such as money earned from past investments, retirement, and savings. A recent article in *American Demographics* stated, "In many non-metropolitan retirement communities, the mailbox is the biggest source of local income."

One community found a way to tap into this new "silver mining." Kremm-ling, Colorado, responded to its aging population by developing an assisted-living center. This created meaningful employment, prevented money from leaving the community, and provided a place for the elderly, thereby keeping the community and its families together.

To paraphrase a statement by Jack Ward Thomas, the former chief of the Forest Service: Not only are economies complex—they are more complex than we'll ever understand. Thomas was referring to ecosystems when he said this, but the same applies to economies. Economic growth is fueled by many factors. An aging population leads to an increase in retirement income. Urban refugees seeking a higher quality of life bring jobs with them, create new jobs for locals, and help drive up the demand for construction, health care, schools, and other services. Economic growth is also related to the phenom-enal growth in the stock market, leading to increased economic prosperity and rising consumer spending. These nonlabor earnings allow for greater mobility, resulting in the growth of communities like St. George, Utah, and Whitefish, Montana. Also, increasing property values in metropolitan areas make rural housing relatively affordable. Add to this the effect of telecommu-nications technology and the growth of footloose producer services, and we see the creation of an urban economy in a rural setting. And, just like cities, this new economy comes with the full array of undesirable side effects, including urban sprawl, pollution, and congestion.

Instead of paying attention to the changing global economy and trying to carve out our own niche, we've allowed ourselves to be distracted, waiting for the old economy to return and fighting each other in the process. What we are losing is the opportunity to make sure we benefit from the new growth. This "urban economy in a rural setting" requires that we put in place incentives to preserve wildlife habitat and agriculture, that we plan our transportation systems more efficiently, and that our kids get the training they need to get the good jobs. This is where our attention ought to be.

The Upshot

Let me offer some suggestions. First, let's stop smokestack chasing, assuming that with tax breaks or other enticements we can convince a large company to move to town. Most new businesses will be small and homegrown. Let's find out what they need, and then develop the physical and fiscal infrastructure that encourages entrepreneurship.

Second, let's find out what it would take for existing businesses to expand and add, say, five new employees each. Do they need access to capital, an educated workforce, and good schools for their kids? If so, then let's redirect money now used to subsidize resource extraction toward education, infrastructure, and start-up capital. If the quality of life, clean air and water, and opportunities to see wildlife are also an important determinant in people's decision to locate in an area, then let's protect these assets.

Finally, let's also be honest about who will likely benefit from a switch to a competitive, global, knowledge-based economy. If there is any consensus among those who toil in the uncertain world of economic development, it is this: We will need a trained and educated workforce, and we will need easy access to large regional and international markets. Communities adjacent to metropolitan areas, or with access to larger markets via an airport or the Internet, will fare better than small, isolated, rural communities that don't have these assets. Communities and states that have invested in education and training will likewise fare better than those who have not.

In my travels I have learned that some are succeeding in this new economy, while a great many are not. But I have also discovered that westerners don't quit easily. We have a tremendous entrepreneurial spirit, and given the right incentives—and the right information—we have always been creative and competitive. A transition to (what is being counted as) a service economy therefore need not scare us. Instead, we should ask ourselves whether we have the education, infrastructure, and political will to capture the high-paying, high-skill jobs.

REFERENCES

Thurow, Lester, *Head to Head: The Coming Economic Battle Among Japan, Europe, and America.* Warner Books, 1993.

Reich, Robert B., *The Work of Nations: Preparing Ourselves for the 21st Century.* New York: Vintage, 1992.

Who Will Be the Gardeners of Eden? Some Questions about the Fabulous New West

Lynn Jungwirth

Essays about the New West and the new amenity-lifestyle-service economy always leave me musing. There is a terrible tension between the New West and the desire for more wildness. Many writers would have us believe we can have both: "You too can be a graphic designer and live on the edge of wilderness." The New West, it seems, wants to be an urban economy in a rural setting.

The current inhabitants of the New West are extolled to understand that their economy is no longer based on natural resource extraction. Service is the name of the new economic game. The economy is really based on the amenities of the area, which draw new money and new businesses. Protect your scenic grandeur, we are told, and you will prosper.

Protection ... a Close-Up View

My community, Hayfork, California, has been busily protecting scenic grandeur since the 1930s when the first forest ranger rode into town. By 1964 we had been asked to set about a third of our 2 million acres into the Trinity Alps and the Yolla Bolla wilderness areas. We did that. In the early seventies we were told of the importance of roadless areas. We agreed to set aside several large chunks of forest and protect them from roads. Then came the Wild and Scenic River designations and the land zoning of the land management plans for the Shasta-Trinity and the Six Rivers national forests. Lately we have protected additional areas through President Clinton's Forest Plan for the recovery of the northern spotted owl. That plan added late seral reserves, key watersheds, and riparian areas to the protected list.

In 1990, U.S. District Court Judge William Dwyer's ruling on habitat for the northern spotted owl shut down the forest to timber management until the Forest Service could produce a recovery plan. In 1994 the judge accepted the president's Forest Plan, and over 85 percent of the 1.8 million acres of federal land in our county were in various types of protection and off-limits to timber management.

We have been uncommonly protective of open space, scenic grandeur, wilderness, and biodiversity. Why is it we are also uncommonly poor? Where are the new economy entrepreneurs and their new businesses? We are surely ready to serve them. Why is it we are not participating in the new prosperity? In 1990 just over 50 percent of our schoolchildren were participating in the federal school lunch program. In 1998 that number was 84 percent. What went wrong?

Wake Up and Smell the Fire

While we were busy alternating between protecting some forestland and managing the rest for industrial forestry, the land under management by the U.S. Forest Service went through profound change. Before the gold miners showed up in the 1850s, the land had been managed by a partnership between the Native Americans, the wildlife, and fire. By 1900 most of those management partners had been moved out of their niches: the miners had nearly destroyed the Native Americans, and understory plants were neglected.

By the 1930s the focus was on the metal in the rivers and the grass under the trees. Thirty thousand head of cattle moved into the area for summer pasture. By the 1950s the gold was gone, most of the cattle were gone, and the focus was on the trees. The understory plants were forgotten. Fire was no longer used to tend the plants. The forest changed.

In 1965 we had our first wake-up call. A 17,000-acre burn raced up Hayfork Summit, quickly moving from grass to brush to trees. The fire crowned, and the forest burned. We were stunned by the loss. Even with aerial seeding the trees did not come back. The south slopes remained in brush fields. The north slopes fared a little better and there were some trees left in the draws the fire had bounced over. But the forest we knew was gone.

In the mid-1970s a fire complex developed after a dry lightning storm in the Klamath Mountains just north of us. In 1987 sixty lightning fires started one afternoon and over 47,000 acres were burned in the Hayfork Ranger District. The burn was very long and very hot, and we lost stand after stand of old growth, second growth, and tree plantations. We were more than stunned. This time we got the message and asked the Forest Service to build a management plan to aggressively address the buildup of fuels in the forest. As I write in late summer 1999, wildfires have wrapped northern California in a pall of smoke. We know now that learning to work with fire in the landscape is the challenge before us.

Fire scientists believe an unusually wet decade around the turn of the century resulted in a flush of seedlings across the mountains of southern Oregon and northern California. By driving out the Indians, the deer, and the fires, we prevented the natural thinning process from occurring. When we cut the old growth, we replaced it with very dense plantations, most of which need thinning. The forest today is that legacy: overstocked and fire-prone, susceptible to fierce, stand-replacing fires.

Tending the Garden

We expected protection of our scenic grandeur to bring both economic prosperity and ecological resilience. We find that has not been true, not for our community and not for our forest. I would like to postulate another version of the New West. In this story we still get to have Paradise Regained, but this time there are gardeners in Eden.

Cee Cee Hedley is a strong young woman with more than twenty-five years of work on public land. She started out as a tree planter back in the 1970s as part of the "Hoe-dads," the tree-planting cooperative in Eugene, Oregon. Today, as an independent contractor, she conducts forest surveys up and down the West Coast. She has planted more than one million trees to reforest western mountains.

Last year I went to a wonderful evening celebration in Washington, D.C., where the conservation community was presenting some national awards. In a lovely speech another young woman presented an award to an older gent from D.C. who had helped create some important legislation or regulation or something to help protect natural areas. She referred to him as a "steward of the land." But all I could see was Cee Cee Hedley with her bagful of a million trees, swinging her hoe-dad, bending down, stepping the tree into place, straightening up, moving uphill, and starting all over again. I knew then that Eden would need gardeners.

Stan Stetson works on the Hayfork Ranger District. For more than twenty-five years Stan has stayed in place, refusing the "you only move up if you agree to move out" career paths of this agency. As a result he has a low GS rating, the kind reserved for people who, in agency parlance, "don't have significant responsibility for making complex decisions." He also has an uncommon knowledge of the land and the people.

This summer the mountains were on fire, and only a few Forest Service folks were left in town at the end of August. It was the time of the Trinity County Fair, the largest social event of the year in this town of 2,500. Stan was at the fair, helping with the mule races, when he learned (through the radio he always wears during fire season) that three small lightning fires had been discovered a short distance from town. He looked through the crowd and found a couple of timber-fallers in the grandstands, then drove through town and found a guy with a water truck and somebody who owned a dozer. Those workers joined forces with the one available Forest Service engine and its sparse crew, and together they put the fires out. Then they went back to the mule races.

No heroes, no awards. These were just everyday activities of somebody who loves the forest and knows his neighbors. No amount of information technology could have done what Stan Stetson did. I knew then that natural resource management *is* a service, perhaps the greatest service of all.

Cee Cee Hedley and Stan Stetson bring a knowledge and ethic to their work even Aldo Leopold would envy. Cee Cee is part of the invisible mobile workforce that has watched the forest landscape of the West change over the last twenty-five years. Stan is the invisible part of the highly visible Forest Service. He has stayed in place, working every day in the forest around Hayfork. The knowledge these two bring to their work cannot be replaced and cannot be replicated through college courses. They live with the land. I believe we can build an economy in the rural West around the knowledge and ethics of folks like Cee Cee and Stan.

Community Regeneration

I have seen this economy in the New West. I have read about it in the *Chronicle of Community* and *High Country News*. Best management practices on cattle ranches have allowed cows and wildlife to prosper. Sustainably managed forests have protected both species and timber production. Profitable ranches and forests have kept green spaces wide open and protected from development. Sustainably produced beef and wood are acknowledged in the marketplace as both a product and a service. The Nature Conservancy is often in the vanguard of this new economy. So are hundreds of ranchers and forest-land owners and workers. This new product/service economy is just as real as the "amenity-based lifestyle" economy, and probably more crucial.

When land is degraded and there is no product to carry the cost of the services, the new natural resource wizards have learned to work with nature to help restore and rehabilitate degraded landscapes. In Eureka, California, the Redwood Community Action Agency created its natural resource division more than twenty years ago. This community development agency fights poverty by teaching people to restore streams in both urban and rural settings. It is considered a leader in watershed restoration design and implementation. The workers are knowledge-rich.

Knowledge-rich forest workers are not a new phenomenon. The art and practice of silviculture is amazingly complex. Foresters have always been well paid. Loggers were well paid until the mid-1980s, when the restructuring of the industry stalled logger wages and increased sawmill wages. Tree planters and reforestation crews, however, have been considered unskilled labor.

In 1994, when "ecosystem management" was the new philosophy for managing public forestlands, we decided to help our workers learn this new knowledge-intensive management and become the high-skilled paraprofessionals needed to carry it out. We started a training class for displaced forest workers, in partnership with the local community college, the local job training provider, the Forest Service, and the Bureau of Land Management. Our local, worker-based, nonprofit group organized and implemented the program, which was structured through three classes: Watershed Assessment, Field Survey and Monitoring, and Field Practices.

Workers attended classes on Fridays and worked Monday through Thursday on forest projects that reinforced their classwork. We now have a cadre of highly trained, highly skilled forest workers ready to assist landowners in managing their forests for biodiversity, clean water and clean air, as well as wood products.

This ecotech training program, or some variation of it, took place in many towns up and down the West Coast. In what seemed like spontaneous regeneration, communities and workers found retraining in ecosystem services a natural and desirable strategy. It was a very small step for tree planters and loggers to see themselves in a new role as stewards of the land. It was also a very small step for communities and agencies to see the value of highly skilled, cross-trained workers. Workers who can help assess the condition of a watershed can also help make good decisions about future management alternatives. Workers who responsibly replace culverts, recontour roads, protect riparian areas, and survey for wildlife and plants can just as responsibly carry out vegetation management objectives through thinnings, fuel breaks, and harvest. The challenge, of course, is obtaining funding for these services.

Companies, such as Collins Pine, that practice certified sustainable forestry have funded these new forest workers through the sale of their lumber. They have demonstrated that it is possible to be responsible for the forest and still generate profit. Public lands, managed just as responsibly, now produce so little product that they not only cannot generate a profit, they cannot even generate enough funds to pay for the management needed to keep the forest healthy.

When will clean water, clean air, biodiversity and scenic grandeur be worth paying for? For the last twenty years they have been worth fighting for. Who decides—and who pays? Will the American voter decide that forests are national treasures and pay for their upkeep? It is the responsibility of those who would control the management of the public forest—the agencies, the vested interests, the public, the industry, and the environmental movement—to recognize the gardeners of Eden, and to make sure they are paid.

Perhaps highly skilled stewards of the land will one day be treated as a collective value. Perhaps a strong partnership between the "amenity-lifestyle-service economy" and the "work-in-the-dirt-and-protect-the-land economy" will give us all what we need: food, shelter, clothing, and a strong and resilient ecosystem. While some of us will benefit from the amenity lifestyle and never get our hands dirty, others are content to be the gardeners of Eden.

The Death of John Wayne and the Rebirth of a Code of the West

Peter R. Decker

Those of us who live and work in the interior West (Idaho, Montana, Wyoming, Utah, Colorado, and Nevada) like to think of ourselves as westerners. We believe we carry with us certain regional characteristics that set us apart from southerners or New Englanders or even midwesterners. In an American society more and more homogenized by modern communications, there is something comforting about a regional consciousness. It gives us a sense of place and identity in a nation where television and the Internet want to erode regional distinctions. Quite aside from our zip codes, we wish to adhere to regional traits and values (some say codes) that we believe are uniquely western. We can't always articulate the components of our regional consciousness, but like beauty or love, we feel it and defend it when others try to fault it or diminish it.

There is, however, considerable confusion as to what these values are, where they came from, and their usefulness in a New West.

Some Myths Die with Their Boots On

What makes it so difficult for westerners and others to discern our regional characteristics is the number of definitions, including myths, that encumber this region. Certainly some of these legends and myths we create for ourselves. But most are externally generated. Some are innocent and benign, causing a few laughs but no injury, while other legends provide us goals and guidance, frequently unattainable in a generation, but worthy objectives to which we aspire. Then there are those myths that degrade and cause misunderstanding—the inaccurate ones, the stereotypes that are sometimes purposefully cruel and cause so much misunderstanding and even violence. Some of the myths die hard, still with their boots on; others refuse to die at all.

Many Europeans, for example, see us as a collection of gun-crazed wild men (and women) impatient to pull the trigger on anyone who disagrees with us—even, sometimes, our children. This habit of violence, some American historians tell us, is a carryover from our recent "frontier" experience. The West for these observers has always been a violent place—our most violent region—one burdened by a "legacy of conquest."

There are those who portray the West as our most democratic and tolerant region, a place where republican principles and procedures flowered in a young region unburdened by European traditions. Because there were few social or economic divisions among the early settlers, it is argued, a pioneer generation passed on to us enduring habits of democratic thought and procedures.

And then there is the Hollywood version of the West as a place where sun-bonneted maidens scamper about the daisies beside their little house on the prairie while Pa, with a six-shooter strapped to his hip, rides his trusty steed out across the wide-open range singing "Don't Fence Me In." A related scenario features the "rugged individualist," the macho John Wayne type, who conquered the wilderness with bravery and true grit while making the West safe not only for democracy but also for Darwinian economics. Tinseltown's most enduring message about the West, one that sells books and films to this day, is revealed in John Ford's 1966 class film *The Man Who Shot Liberty Valence.* When Jimmy Stewart's character in the film finally reveals to a newspaper editor the ironic truth of who really shot the notorious outlaw Liberty Valence, the editor refuses to rewrite the story and proclaims, "This is the West, sir. When legend becomes fact, print the legend."

In addition, there are those who would have us believe we are a plundered province, aided and abetted by the West's own residents who care nothing about the environment and everything about money. We are told we've overgrazed, overcut, and overmined what was once a beautiful place, leaving in our wake a raped landscape with dusty pastures, barren hillsides, and poisoned streams. In short, we're poor stewards of the physical beauty that once surrounded us.

The problem with all these myths, of course, is that they fail to account for history. It is difficult to defend the West's monopoly on violence, for example, when looking at the Civil War and the Ku Klux Klan in the South. As for our heritage of regional tolerance, clearly Hispanics and Indians question this myth. And did not the Sioux teach Custer the deadly lesson that rugged individualists cannot survive long in either war or the West? As for ruining our environment, we are certainly not alone.

A more enduring myth, one put upon us by an eastern cultural establishment, would have us believe that the West is something of an intellectual outback, lacking for world-class museums and educational institutions, a region where "minor regional writers" struggle, like the West itself, with their own self-identity . . . and the composition of a compound sentence. We're viewed as a collection of provincial dirtbags, too lazy to move to more sophisticated and cosmopolitan regions of the country. All the energetic and smart folks born and raised in the West departed long ago, leaving behind a region described by one eastern sociologist as "a fished out pond . . . full of bullheads and suckers."

This view was best expressed in a recent review of a new novel (*Plainsong*, by Kent Haruf) about the lives of decent, hardworking, and caring people on the plains of eastern Colorado who look after each other in tough times in a tough country. The reviewer, Joyce Carol Oates—a well-respected writer and teacher of creative writing at Princeton—can barely hide her contempt for these rural residents. She finds them "boorish," "lacking in imagination," people who live in a "narrow, ignorant, smugly provincial domestic world," and who are incapable of "embarking upon the adventure of independence in Chicago, and beyond." What is worse, this critic continues, these rural Coloradans "live not in the solitude of their own thoughts, at the margins of the community, but in one another's lives." Damned dangerous folks, these hayseeds. Sounds like they might be acting like neighbors in a community! People just don't act that way anymore, the professor tells us, and anyone who writes about the older, traditional values of rural life is engaged in an exercise in "fantasy." What we need instead, we are told, are more books that portray the dark side of rural life, books like William Faulkner's *The Hamlet,* which features a speechless idiot and his liaison with a cow.

When it comes to understanding the West and rural life, which is so much a part of understanding this region, Oates is quite clueless. She needs to understand that the pastoral life does exist for some people in the West (as elsewhere) and that life experiences in rural communities are not fantasies; that westerners are not a rag-tag collection of demonic characters living and working out in the Big Empty who wish to be described and understood by writers whose specialty is "metaphysical speculation."

It is one thing to read and listen to these stereotypes as they bombard us from other regions, but some of the voices are also coming from the thousands of new residents moving to this region and the millions of tourists vacationing here. They perceive the region as a place to vacation and recreate—a sort of fantasy theme park where cowboys and Indians define the workforce and mountains and rivers define the workplace.

Defining a Code of the West

If these are some of the more destructive stereotypes and myths that derive from our mistaken identity, what then are the codes by which we live here in the West, and from where do they derive? Let me suggest one major source and a few historical values that, though often forgotten, flow from this source and continue to survive in today's fractured West.

A *Working Landscape*

What is unusual about the West, what makes it so unique as an American region, and what to large degree defines its people (culture), is our landscape. It is one as immense in scale as it is varied in composition—a complex web of forests, mountains, rivers, deserts, and grasslands. It is also a landscape that over time and throughout its space has experienced human contact—more gentle contact, certainly, prior to the late nineteenth century than recently, but contact with people nevertheless.

The term "landscape" has always carried within its definition the presence of people, living and working on the land. Cultural historian Simon Schama refers to it as a "manscape," a place of human habitation. Even Henry David Thoreau, that misanthrope naturalist, recognized the historic connection between Walden Pond's wild landscape and the local Indians. Only recently have we in the United States and Europe begun to think of the landscape as an aesthetic, a place of pastoral beauty where people and their work are either totally absent or at least barely visible. It is this massive and varied western landscape—our interaction with it and our life and work on it, generation after generation—that defines the contours of our western culture, our memory and our history. And more than in any other region of the country, it is the landscape that influences western behavior and helps to define our most important codes.

This is true, I think, because the West, more than any region, is not far removed in time from the soil and its agricultural heritage. Hispanic migration into the interior West occurred little more than two centuries ago, and Anglo settlement is even more recent. The descendants of these settlers and thousands of others who today continue to live on the land, work its fields, graze its meadows, and harvest its crops do not consider the landscape as an abstraction, a recreational playground, or a commodity to be traded and monetized. People who make their living from the land see it as a workplace, the source of their livelihood and the place of their home. And although the traditions that derive from the land are not uniformly present throughout the West, particularly in our cities, the landscape continues to be the primary provider of memories and self-identity for many rural westerners.

A Region Built on Cooperation

It is this agricultural experience, this intimate connection with our landscape, that provides us with many of our codes. As a region, we place a high premium on hard work, honesty, thrift, and tenacity, and to work the land involves all four qualities. To this day, the work ethic remains strong in the West. And because our relative wealth did not come easily, we are careful how we, and others on our behalf, spend it. Nor are westerners quitters in the face of overwhelming odds. The close relationship to our agricultural heritage, however, involves something else, a code more important and enduring.

As I've suggested, settlers did not come to this place alone nor could they survive as rugged individualists. Those who migrated into the West along the trails and rivers from the East and the South did so in highly organized immigrant companies, often resembling military units. And upon arrival on the high plains or in the mountain valleys of the Rockies, settlers understood quickly that any attempt to confront the forces of nature alone in this dry, harsh country was an invitation to failure, if not death. Survival depended on hard work and patience, but it also depended even more on cooperation. Hardship and adversity only strengthened the custom.

In a new region where there was little capital to purchase labor and no excess labor to supply the need, the exchange of labor and scarce tools became a habit of survival. In this spirit, people together built houses, barns, roads, irrigation ditches, fences, and a host of institutional structures such as churches, schools, firehouses, courthouses, and jails. In addition, a series of economic recessions, and particularly the Great Depression, which hit harder and longer in the interior West than in any other section of the country, embedded in the survivors a wide array of skills and memories that survive in their children and grandchildren to this day. Those who managed to stay on the land (and fewer than a third did) reminded themselves and their offspring of life's harsh lessons: avoid unnecessary risks, especially debt; work hard; pray for God's grace; hope for a dividend of luck; and, above all, cooperate with your neighbors in good times and bad, especially bad. Imbued with a new, grimmer view of life's possibilities, no one expected to be cut any slack, now or in the future. And if these survivors felt like passive agents in a national economic system they neither understood nor profited from, they had at least survived with one another's assistance. And that was important.

This code of cooperation carries forward to this day. Farmers and ranchers share labor and equipment, and together they work brandings, roundups, and harvests. Rural people call it "neighboring," a verb that simply translated means helping others. It can be something as simple as repairing an irrigation ditch or as important as helping to care for a sick neighbor. A good neighbor will return the assistance, the exchange of labor, out of personal friendship and a shared sense of community responsibility. No one keeps an accounting of how much labor is given and how much is received quantitatively, because no one is willing to equate or reduce personal relationships to a definitive value—not because it can't be done by some sophisticated mental calculus, but because it is inappropriate, even rude, to even attempt such a calculation. You are just brought up to learn that labor and assistance are mutually shared, in good times and bad.

"Neighboring" includes also a high degree of mutual trust and predictable behavior where a promise made, is a promise kept. Not to do so is to destroy the trust upon which the entire "neighboring" edifice is built. Let me cite one personal example.

Some years ago, after a very dry summer, I sold 100 tons of hay to a local rancher. My neighbor said he'd pay me half now and the other half in the spring when he picked up the hay. We shook hands on the arrangement. As agreed, my neighbor picked up his hay in March, except the check for the balance due was more than I expected. "Why the difference?" I asked. He replied, "Oh, the bales weighed a bit more than you thought . . . total came to 109 tons, so that's the difference." A year and a half later, in the midst of an IRS audit, the agent asked to see the contract for this hay sale. "There is no contract," I responded. The IRS agent looked at me as if I had just fallen off the onion truck. "No contract for a $7,000 sale?" "That's right," I informed him. "We don't use contracts around here, at least not with neighbors."

I suggested that if he didn't believe me (which he didn't) he should call my neighbor (which he did). When I visited with my neighbor to apologize for the inconvenience of the IRS intrusion, I told him my audit occurred on the occasion of my daughter's fourth birthday party. It was, I believe, the first time an IRS agent had ever been made to play Pin the Tail on the Donkey. My neighbor thought a moment and responded with a grin, "I sure hope you used the IRS ass as the object."

The "rugged individualist" doesn't get any of this, of course. He represents the individualism of the self, the self-congratulatory braggart, who doesn't understand that a mature independence carries with it the assumption of responsibility to a larger community. The "rugged individualist" proudly protects his privacy in his urban bunker or on his ranch compound, both posted with "No Trespassing" warnings. With his large bank balance, he smugly assumes he doesn't need—nor will he be indebted to—his neighbor's assistance. He can, of course, buy all of his ranch labor and the legal talent needed to prosecute those who violate his "property rights." In his arrogance he will go it alone because that's the way he made his money. He doesn't realize, at least at first, that the practices on Wall Street are far different from the codes on Main Street.

The Importance of Open Spaces

There is, I believe, one other important quality of the West. I hesitate to call it a personal code; it is a physical characteristic that influences our behavior and, to a degree, places a premium on the code of cooperation. I'm referring to the immense space that surrounds us. The sheer size of this space—all 100,000 square miles of it—affects our psyche. As westerners, we carry within our mental geography a different, larger sense of space, distinct from New Englanders and certainly from city residents. We are accustomed to more space surrounding us, and therefore we need more elbow room to feel comfortable, and we get damned grumpy when newcomers crowd us—on a highway or in a new subdivision across the fence. It is all this gorgeous open space that now attracts others to this region.

We find it difficult to absorb these residents into our communities. As newcomers crowd in around us with their different sense of space and place, they want to question the established norms and traditions in their new surrounding; they want to lecture us on a multitude of subjects, especially how to reorganize our landscape. It is on our landscape, we say to ourselves, they wish to relax and vacation, the same landscape on which we wish to work and harvest. We are accustomed to more space and fewer structures, more acreage and fewer roads, longer sight lines with more sky. Newcomers do not always share the same vision. I'm not suggesting our view is any better; it is just different. And it is these differences that have caused considerable internal debate and raised the noise level, if not the contentiousness, in our western cities and towns.

Maybe it is time westerners remember one of our strongest traditions, that of being a good neighbor. This western code of "neighboring" has always assumed a degree of tolerance and a willingness to listen to different opinions. And there is, of course, a mutual obligation for others to do the same. Being a good neighbor includes also taking care of our landscape and being good environmentalists. Agriculturalists and others who have lived on our landscape possess an authentic knowledge of what works on the land and have learned, sometimes the hard way, that to consume our landscape is to destroy our livelihood. If all of us, natives and newcomers, cannot cooperate, how can we possibly expect to preserve the western landscape we as a nation so cherish? To do anything less is to be a poor steward; but to do more is to be a good neighbor.

It is easy, in these cynical times, to be charged with nostalgia. But if nostalgia is looking to our past for something that worked, a code that allowed people to live a life of dignity, individually and within a community, then I plead guilty. If this is nostalgia, maybe we should have more of it . . . and let it blossom throughout the land.

What Is Community?

Carl M. Moore

We live in an age in which people have become increasingly individualistic—putting their own growth and advancement ahead of almost all other considerations—but we find, paradoxically, that we need others to find ourselves. It is through community—the meaningful interaction with others whom we know—that we make sense of our own lives. "Community" once meant growing up in a place where children didn't dare get into trouble during the week because they would hear about it when they went shopping with their mother on Friday at the A & P. It was a time and place and a set of relationships that differ profoundly from so many of today's urban and suburban neighborhoods.

People concern themselves with community for many reasons. The first is most basic. People gather together instinctively. We live according to social norms and occasionally we change them. We are curious about the social forms we live in, about why they exist, and about whether they serve us well. The discipline of sociology was created to explain the changes that were taking place in society as a result of the Industrial Revolution. Most classical sociologists—Max Weber, Emile Durkheim, Karl Marx, Georg Simmel—have given significant attention to the theme of community in light of the changes they observed in social patterns.

The changes stemming from the Industrial Revolution, particularly the need for a concentration of people to work in mechanized industries, resulted in the dislocation and isolation of many. The seeming paradox discovered by early sociologists and psychologists was that the larger the population, the more people became isolated as a way to protect themselves. Two of the most marked changes in society are the evolution away from family (extended kinship groups) and the migration away from the traditional village to the city. Driven in upon themselves, people seek environments that are more "communal." They are acting to regain something that has been lost because they do not want to end up naked and alone.

Neighborhoods were created within cities to enable people to live in smaller—more community-like—environments. People built towns and "suburbs" (sub-urban) that were close by but apart from the city. We have invented living patterns that attempt to create community in the shadow of the city.

In addition to becoming increasingly individualistic—producing isolation, separation, and alienation—some of the most perplexing obstacles to community include the following:

- Intense nationalism—and related "isms": racism, ethnocentrism, anti-Semitism—yielding hatred and a strong tendency to "scapegoat" outside groups.
- Widespread reliance on bureaucratic social structures, stressing hierarchical distinctions, specialization, and centralized authority.
- Rampant consumerism, with its emphasis on the accumulation of things over the pursuit of interpersonal connections.
- Popular indulgence in the fads promoted by mass media.
- The fast pace of life, in which individuals rush about to fill their lives with activity.
- Emphasis on professional success above all other considerations.

For the past three hundred years, approximately since the time of Descartes, people have believed that it was possible to discover rational solutions for any problem. But we have come to realize that complex human problems cannot be solved by rational thought alone. Experts do not know enough, can never know enough, and, ultimately, are not responsible for the outcomes that follow from their advice. It is the people *with* the problems who must determine what solution is right for them. They can benefit from expert advice, of course, but that is an insufficient basis for making decisions. We must learn ways to bridge our differences so that we can work together to solve problems in our own communities.

What people throughout the country sense they have lost, and want to regain, suggests the key elements of community. Keep in mind that a definition of a sociological concept like community is going to be an "ideal type," an abstract idea that can never be achieved as defined. Put another way, think of community as Plato thought about art. The ideal community would be in the demiurge. To draw a picture of it, on canvas or on the page, is to make a copy, never quite capturing the actual thing. Moreover, since community is in the struggle for community, no place gets it just right, or there would not be community.

Despite these caveats, here's a working definition: A community is the means by which people live together. Communities enable people to protect themselves and to acquire the resources that provide for their needs. Communities provide intellectual, moral, and social values that give purpose to survival. Community members share an identity, speak a common language, agree upon role definitions, share common values, assume some permanent membership status, and understand the social boundaries within which they operate.

A community is larger than the most personal components of a society, such as couples, groups, families (even extended families), but smaller than the most complex components of a society, such as a large city or a region or a state. Community exists when people who are interdependent struggle with the traditions that bind them and the interests that separate them so that they can realize a future that is an improvement on the present.

All communal forms have a political nature. For any collection of persons to live together over time there must be an ultimate appeal to some kind of finality, to authority or power. As Judith Martin and Gunther Stent so aptly put it in their essay in *American Scholar* (Spring 1990): "There can be no such thing as civilized living in the absence of etiquette and law. Even if one has a well-developed intuitive feeling for the moral point of view and manners, one cannot navigate through civilized society by social instinct alone, or by mere reliance on one's human nature."

Commitment to a community is likely to exist if there is a communal return, if people derive a sense of belonging, recognition, or acceptance from being part of the community. Communal membership must be satisfying on many levels of experience and must involve emotional and physical investment and returns. One form of communal return is that community provides the "stage" upon which the individual may achieve integration. Community is the context in which the person is viewed as complete.

Conflict is essential in creating and re-creating community. If there were no differences among people, there could be no community. But differences alone do not make a community. Community is forged out of a struggle by people to determine how they can live together. One of the critical requirements of any community is to invent the processes of interaction that allow people to live together.

There are some groupings within our culture that are often referred to as communities but that do not, in fact, meet one or more of the important criteria implied in this definition. These groups may lack identity, commitment, commonality of place, differences, or the motivation to struggle with their differences.

Obviously, some suburbs meet the definition of an authentic community. Most of them, however, are not ever likely to be a community because the people who live there have little or nothing to do with their fellow residents other than sharing a common lifestyle; their concern for the well-being of the community is likely to be limited to concern for the retention of property values; and residents are not likely to wrestle over tough issues with people who are very different from them. Even where there is a high level of involvement in a suburb, its citizens are not likely to develop their problem-solving capacity. They do not need to. There are no perceived real differences to overcome. There is nothing local that concerns them enough to mobilize. Their real struggles are likely to occur over economic and work issues, and those are likely to take place within the work environment.

It is in the nature of a cult to be insular, to be intolerant of differences. A cult is in effect a community with walls, and it fails as a community for the same reason—not so much for what is walled in but because of what is walled out. On the other hand, coalitions are held together only by interests, not by shared values and culture.

Mediated communities are connected by some electronic means. They are not defined territorially and members may never come together face-to-face; the electronic medium is their commons. Such groupings include electronic networks connected for the purposes of work, play, or shared interests. These groups may eventually meet the definition of an authentic community if they develop traditions that join the individuals within them in meaningful ways, and once the mediating medium is able to facilitate effective conflict resolution. Given their current status and sophistication, they do not meet my definition of community.

Professional communities—whether defined broadly, as "the scientific community," or narrowly, as a particular academic community—exist only because of their shared interests. Organizations and institutions can have characteristics of a community, but they are not likely to be communities.

Intentional or designed communities do not become communities because of their design. In fact, most such experiments reveal the limits of planning and the physical environment in determining what is and is not a community. The story of Levittown, Pennsylvania, probably the first completely planned suburban township, reinforces the point that community emerges only when the people begin to struggle with their differences.

Moral collectives or religious groupings typically are referred to as communities. While they do meet some important criteria—the people are interdependent in many ways, share traditions, identify with each other, and share a common commitment—they usually do not struggle over hard problems with people who are truly different. Eleven o'clock Sunday morning is the most segregated hour in America.

Finally, it is not realistic or useful to assume there can be a total community to which we all belong, or that we are all participants in the global community. While the spirit of the expression is understood and appreciated, the reality is that if the term community is to have any utility, it cannot be used so cavalierly.

What is the benefit of people being armed with a definition of community? If someone is interested in change and can get others to share in a common understanding of what community is and is not, they have permission to ask whether the citizens of their community are struggling with the interests that separate them. If not, what can be done to get them involved? Are there processes in place that can be used? Or do new approaches for how to struggle together need to be invented or borrowed from other places that seem to be working more effectively?

A definition of community such as the one presented here encourages people to find creative ways to celebrate the traditions that bind them together in the neighborhood. Even when times are hard, there is a civic infrastructure when people are willing to grapple together over tough issues. Interdependent people, struggling with the traditions that enable them to come together and with the interests that separate them, can realize a different future.

ON THE GROUND:
COLLABORATIVE CONSERVATION
IN PRACTICE

Skeptics of collaborative conservation (and there are many) complain that the many watershed groups, resource councils, and partnerships add up to little more than the sum of their meetings—in other words, that collaborative conservation has failed to show positive results on the ground.

Our experience is otherwise. Each of us has participated directly in the formation of collaborative groups that accomplished measurable environmental benefits; we have observed and heard about dozens of others in the course of our research. We have also participated in, observed, and heard about many other initiatives that haven't gone beyond the "getting to know you" stage, or have ended in discord of one sort or another. We believe the lessons from the latter category are equally important as those of the successful examples.

This section tells the stories of eight collaborative initiatives in the West. We lead with the Quincy Library Group, whose legislative end run around the Forest Service has garnered more press about the collaborative conservation movement than any other initiative. Although we believe Quincy may be a poor choice as a poster child for collaboration, it has become a touchstone in almost all discussions of the subject and thus is an important example of the ways such groups have organized and acted to implement their goals.

The subsequent stories in this section are of quieter but equally important initiatives: the farmers and environmentalists who overcame substantial cultural barriers to hammer out a management plan for Montana's Clark Fork River; the community leaders, timber industry officials, and environmentalists who convinced the Forest Service to embark on innovative experiments in the Applegate Valley in Oregon; the ranchers who banded together in an effort to preserve their open lands and work cooperatively with public agency officials and private land conservation groups in the Malpai Borderlands region of New Mexico and Arizona; and the residents of Routt County, Colorado, who approved a tax increase to finance the purchase of agricultural conservation easements and thus preserve ranchlands around Steamboat Springs.

We also present the story of artists in Washington's Olympic Peninsula who helped their community celebrate and understand rapidly declining wild salmon runs. Their unconventional methods—with more than a little humor, sentimentality, and adventure thrown in—have strengthened community spirit as well as salmon populations. For its part, the story of Oregon's ambitious salmon recovery plan offers lessons in the political and legal realities of tackling such problems on a larger scale.

Finally, we offer what can only be described as an unfinished story, an account of a proposed Citizen Management Committee for reintroduced grizzlies in the Selway-Bitterroot mountains of Montana and Idaho. Met with intense opposition from both ends of the political/interest group spectrum, this collaborative group's experience well illustrates the perils of staking one's place in the "radical middle."

The Quincy Library Group:
A Divisive Attempt at Peace

Ed Marston

Perhaps the best-known example of a collaborative conservation initiative, the Quincy Library Group (QLG), is made up of timber people from California's northern Sierra Nevada, local environmentalists, citizens, and representatives of local government. It had been just another consensus-based partnership until 1997, when U.S. Representative Wally Herger (R-CA) introduced a bill ordering the Forest Service to do the group's bidding. After a contentious House debate in July 1997, the bill passed that body 429 to 1. Western Republicans such as Alaska's Don Young and Idaho's Helen Chenoweth joined enthusiastically with green Democrats Vic Fazio of California, Peter DeFazio of Oregon, and Republican environmental leader Sherwood Boehlert of New York to push the bill. Finally, even a very reluctant George Miller (D-CA) threw his support to the bill, bringing a substantial bloc with him. The proposal passed the Senate the following year as a rider to the 1999 spending bill, after facing fiercer opposition, and was signed into law on October 22, 1998.

The bill tells three national forest supervisors in the Sierra Nevada Range to revise their forest plans to incorporate the group's proposals in over 4,000 square miles (2.5 million acres) of national forest—a land area twice as big as Yellowstone National Park. The project aims to log about 9,000 acres per year of roaded, second-growth forests, and thin another 40,000 to 60,000 roaded acres annually to protect against fire. It will also put 150,000 acres of roadless,

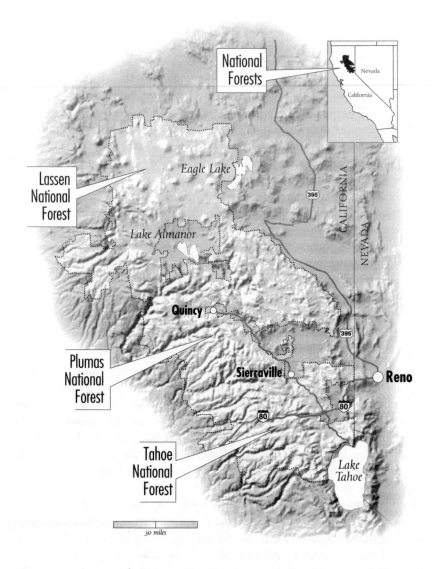

old-growth land off-limits to logging, protect all trees over 30 inches in diameter, and provide extrawide buffer zones against streams.

Environmentalists outside of Quincy have reacted negatively. They like parts of the plan, but object to its expense, the money it might pull from other forests, the greatly increased logging, and the geographic scale. Most of all, they object to passing laws to manage individual forests—especially laws written by what they see as a local, narrowly based, special-interest group.

"Working-Class Gorgeous"

There is nothing about Quincy, population 5,000, to tell a casual observer that the environmental movement's most serious challenge in decades started here. Quincy looks like the logging town it has been since the turn of the century. Trees grow thickly right up to the back of house lots, and although the town is ninety minutes north and west of Lake Tahoe, and four hours north and east of the Bay Area, it still looks rural.

But change is coming, observes the QLG's lead environmentalist, attorney Michael Jackson, who points to the town's two coffee shops: "Bob's Fine Food closes at 9:00 A.M. because their customers have gone to work. It has customers starting at 4:30 A.M., when we newcomers (Jackson has been in Quincy since 1977) are still asleep." Bob's serves the kind of coffee that comes in round cans; the other shop, Morning Thunder, opens later in the morning and serves cappuccino and espresso.

Though Quincy has growing numbers of new people, this is not the Sierra Nevada that inspired the creation of the Sierra Club. QLG member Linda Blum calls it "working-class gorgeous." A climb takes you to the tops of softly rounded mountains covered with trees, rather than into the tundra and peaks that characterize the Sierra Nevada Range south of here. Quincy is not as rough or sprawling as some of its neighboring towns, but neither is it busted. California lost sixty timber mills in the last decade, according to the California Forestry Association. Quincy has one of the sixty-one survivors: a huge Sierra Pacific Industries mill now retooled to cut small timber. From Monday to Friday, Quincy's two one-way main streets are dominated by logging trucks on their way to and from that mill. But these are not the good old days of the 1980s. Today each truck off public land is likely to carry ten to twenty relatively small logs; before the California spotted owl, trucks would lug their way through town with only one to three logs of mammoth size. In those days, Jackson said in an interview, when a truck carried a particularly large log, the driver might drive around the block that had his office on it, taunting the environmental litigator with the sight of another fallen giant, before stopping for a celebratory beer at a bar on the town's main street. There's another difference, Jackson said: "These days, when people wave at me, they use all five fingers." Jackson and his allies gave as good as they got during the timber wars. "We blamed and ridiculed our neighbors. There was sugar in the tanks of logging equipment."

Not Conventional Consensus

If "consensus" conveys a smiley-faced image of people sitting around a table making sure everyone is in agreement, the Quincy Library Group isn't consensus. For starters, not everyone is at the table. At a meeting held after the House victory but before the Senate battle, a few Forest Service people sat silently on chairs against the wall, while the group's twenty or so members sat at a central table. Also missing were national and regional environmental groups, who had achieved a consensus of their own. One hundred and forty groups—some from as far away as Alabama and North Carolina, as well as thirty-five California groups and western regional groups like the Southern Utah Wilderness Alliance and the Pacific Rivers Council—opposed the QLG bill. Finally, the group is very definitely led: by Jackson; by former Plumas County supervisor Bill Coates, who resigned his elected position to become a coach at the local Feather River Community College; and by Tom Nelson, chief forester at Sierra Pacific Industries, which owns the Quincy mill and other California mills.

Jackson, the group's tactician and its major representative to the outside world, describes himself as practicing "catch and release" law. He sues environmental agencies for breaking environmental laws, collects his fees from them, and then lets them go, to bring him another fee at some later time. The other half of his time, he says, goes to the Quincy Library Group, for which he does not get paid: "It's the most expensive project I've ever been involved with."

As Plumas County supervisor, Coates used to spend a fair amount of time flying to Washington, D.C., to plead for bigger cuts off the local national forest—his county of 20,000 has depended on federal timber receipts as well as on local taxes paid by the timber industry. Since founding the QLG, though, he began flying to D.C. with his former opponents to lobby on behalf of the group's positions.

Initially, the relationships were uneasy. The spouse of one environmental member of the group recalls that she used to cross the street to avoid Coates. The third leader, Nelson, works for the largest private landowner in California. Sierra Pacific Industries, with 1.4 million acres, is California's Weyerhaeuser or Plum Creek in terms of scale.

Creative Environmentalists...

The Quincy Library Group is a direct result of the creativity and foresight of a relatively small number of local environmentalists, banded together as the Friends of the Plumas Wilderness. In the mid-1980s, when huge logs were pouring off the local forests, the beleaguered greens designed an alternative to cutting old-growth trees. According to Jackson, the approach was put together with research help from The Wilderness Society and forest economist Randal O'Toole, legal help from the Natural Resources Defense Council, and, in the appeals and litigation stage, with the Sierra Club as plaintiff.

The 1986 proposal called for putting all remaining roadless areas into a preserve, with logging restricted to roaded, second-growth land. Mike Yost, a forestry professor at the local Feather River College, recalls the computer and map work that led to the plan: "We basically protected all the land we could—streams, roadless area, old growth—and then told the computer to thin the remaining [roaded] forest heavily. We were trying to show there was volume out there."

They proposed doing the thinning with group selection, which Yost describes as "clear-cuts from a tenth of an acre to a couple of acres." These are small enough to be reseeded by the surrounding older trees, leading eventually to an uneven-aged forest. At the time, Yost said, the Forest Service was using "40-acre clear-cuts followed by herbicide spraying." The Plumas National Forest considered the alternative as part of its forest plan revision, then rejected it in favor of business as usual. But the agency's adopted plan did not get very far. Even though George Bush was in the White House, timber sales began to decline in the late 1980s. Jackson says the forest plans were frauds, written as if Congress were going to pour money on the forests even though the agency knew it wouldn't.

When federal timber stopped coming off the forests, and mills in the surrounding area began to close, Quincy got hot. Sierra Pacific Industries closed its mill during public hearings so that angry workers could attend, storeowners put up yellow ribbons in their windows as a sign of solidarity with timber, and "things got scary," says Yost.

Jackson says, "The environmental community wasn't going to be intimidated, so we were filing appeals on every timber sale."

Then came the owl. In 1993, the Forest Service imposed regulations to protect the California spotted owl, a relative of the northern spotted owl. Linda Blum says the California spotted owl regulations (CASPO, in the jargon) meant the end of business as usual on the local forests: no clear-cuts, no even-aged management of forests, no cutting trees over 30 inches in diameter (old-growth trees can be 7 feet in diameter), and no cuts that opened up too much of the forest canopy. Even before CASPO, some in the Sierra Nevada timber industry could see that the old game was over, and that maybe the environmentalists' 1986 proposal hadn't been so bad. In fall 1992, Plumas County supervisor Bill Coates put in a call to his political archenemy, Michael Jackson.

As Jackson tells it, "Mr. Coates said, 'All right, we're through. We've got to do something new. Will you meet with the mill owners?'"

That wasn't an easy question. After twelve years of Reagan-Bush policy that had wreaked havoc on the forests, logging was nearly shut down. Moreover, national politics had just taken a sharp turn: Clinton was president-elect and the future looked green. But healing wounds in the community was also attractive, especially if industry was willing to agree to the 1986 alternative. Jackson and his allies signed on for at least a short ride. The ride started with "three meetings in the back room of my office," says Jackson. They were supposed to be secret. But, Jackson says, "it was clear to everyone that something was going on, and [secrecy] wasn't acceptable to anyone."

So the next meeting, he says, was a six-hour one in the Quincy Library, where everyone could see it and listen in. (One myth says they met in the library so that no one could shout; another says they met there because local timber people were fearful of being seen going into Jackson's office; a third says the library had a room available.)

In the spring of 1993, the conspirators called a meeting for the public and 250 people came to the Town Hall Theatre. After explaining what was afoot, the leaders asked for a straw vote. Jackson says, "Five wanted to stay the way we were; the rest wanted to move ahead." Two of the opponents were Jackson's, Blum's, and Yost's old allies at the Friends of the Plumas Wilderness; three were with the timber industry.

Jackson, who is not always gentle with his rhetoric, calls the people on the fringe "wing nuts," and says they play the role of keeping the center together. With what it took as a public imprimatur, the QLG went to work, and in June 1993 published its Community Stability Proposal, based on the environmentalists' 1986 plan. Its most important act was implicit: It laid claim to 2.5 million acres of Sierra Nevada public land—all of the Plumas and Lassen National Forests and the Sierraville Ranger District of the Tahoe National Forest. The plan had something for everybody. Industry would get wood from 1.6 million acres of managed land (another 1 million acres were in various reserves, including wilderness), and communities would get jobs and fire protection. Environmentalists would get preservation of 148,000 acres of roadless, old-growth land that was, and still is, part of the three forests' timber base. They would also get expanded protection of riparian zones and an end to 40-acre clear-cuts in favor of small openings, less than 2 acres in size, in the forest. The plan also had aesthetic and landscape goals. QLG members argue that small-patch logging and thinning will eventually return at least the drier parts of the forests to the open, brush-free state, dominated by large trees, that the first settlers found. Jackson says of the bargaining that led to the QLG approach: "The only thing I gave up was my prejudices."

SNEP, CASPO, and All That

An agreement in principle is one thing. Translating it onto the ground is quite another. So the group, Blum says, set about educating itself by bringing in scientists: the ones working on the (still not finished) California spotted owl environmental impact statement (EIS), the ones working on the $6.8 million

Sierra Nevada Ecosystem Project (published in 1996), and well-known experts like Jerry Franklin of the University of Washington and Norm Johnson of Oregon State University. The group also spent a lot of time with certain Forest Service specialists; its approach to fire was taken from the work of Bob Olson, fire specialist on the Lassen National Forest. Once the group had a sense of how to implement its principles, the members asked the three foresters to amend their forest plans to achieve its goals. Michael Kossow, who runs a water-quality consulting firm out of Taylorsville, a wide-spot town near Quincy, remembers a February 1994 trip to Washington to talk to top Forest Service and administration officials: "When we went to D.C. three years ago, [then Forest Service chief] Jack Ward Thomas told all three forest supervisors to go back and start forest plan amendments" to examine and possibly adopt the QLG's ideas. "I was sitting right there. I heard it."

But over the next two years, nothing happened, Kossow says, and that led to the groups' antiagency stance. "I think they got frustrated. They expected things to happen." Kossow also says, "We knew that Thomas met with the supervisors alone the next day. It may be that he told them something different at that time."

High Country News reporter Ray Ring interviewed Jack Ward Thomas in the spring of 1997 in Missoula, Montana, after Thomas had retired from the Forest Service. He told Ring that he disliked almost everything about the Quincy Library Group, especially the fact that Sierra Pacific Industry was involved, and that Thomas's political boss, Secretary of Agriculture Dan Glickman, was backing it. Thomas said, "[Red] Emmerson, who owns Sierra Pacific, is involved in the Quincy Library Group, which is patently illegal. They're a nonprofit charter, and they're violating, not following, the forest plan. But the secretary of agriculture just turns around and says: 'OK, you guys can manage these two and a half national forests.' They're not properly chartered, and they're sitting there cutting deals back and forth . . . I like cooperation, but I don't like Emmerson; who the hell turned over my national forests to him?"

The Quincy Library Group members didn't know Thomas's feelings, but they could see that the local forests weren't implementing their plan fast enough to suit them. So in 1996, members began flying back and forth to Washington, D.C., to lobby for a bill. The Wilderness Society, the Natural Resources Defense Council, and various California environmental groups had been keeping a distant eye on the QLG, mainly by talking with Jackson and Blum. But the shift from lobbying the Forest Service to lobbying Congress got the national groups' full attention. They saw the QLG setting a precedent of piecemeal national legislation for individual forests that would inevitably lead to the overriding of environmental laws.

So in early 1997, a series of tense and often angry negotiations, moderated by Undersecretary of Agriculture Jim Lyons, took place in California between the environmental members of the QLG and the representatives of national and California environmental groups. The non-QLG environmentalists wanted a smaller geographic scale, a slower pace, and—above all—no legislation. Linda Blum says, "We resisted that. We'd been waiting for four years for the administrative solution, and had gotten nothing but excuses and obfuscation and sabotage by various persons working for the U.S. Forest Service at various levels."

Louis Blumberg (who, until 1999, represented The Wilderness Society in San Francisco) was not impressed by the QLG's impatience: "We have all spent many years waiting for the Forest Service to make changes. It's slow to change, but to say that because the QLG is unable to have their plan adopted within two years is testimony to a failure of forest planning—I can't buy that. They're only one special-interest group." He also said that the three forests have been implementing the QLG's program. "Last year, 40 percent of the project was done." Blumberg said that the QLG plan would double logging on the three forests, to 250 million board-feet, and that the estimated $83 million the plan would cost would take away scarce resources from other national forests.

In addition to the threatening precedent of overturning normal forest planning, Blumberg said, there's the scale: "The Quincy Library Group's bill covers one-third of the Sierra Nevada." He argued that the bill will fragment any attempt to manage the Sierra Nevada Range in a holistic way. Jackson understands the fear of setting a precedent. He says the timber industry people on the QLG, as well as western Republicans in the House, feared the precedent of outlawing clear-cutting and entry into roadless areas, while strengthening riparian protection.

The Plan

The Quincy Library Group's approach is audacious. It is a pilot project that aims to return almost 3 percent of California to what it calls a presettlement state. Back in 1908, when the just-formed Lassen National Forest surveyed its domain, 70 percent of the land was in trees over 30 inches. Today, only 13 percent is. Instead of the 20 to 30 old and large trees that dominated an acre in the drier parts of the 1908 forest, much of today's forest consists of a thicket of 1,000 or more small trees per acre, with dead ones on the ground or leaning against the living trees. In the past, fire would have periodically pruned the dead and smaller trees, the group argues, leaving the way open for a small number of survivors to grow large and old. But fire suppression and overgrazing have stopped the ten or so low-intensity fires that would have moved through the drier parts of the forest this century, and the three or so fires that would have struck the wetter, higher, more western parts of the forest, according to fire specialist Bob Olson of the Lassen. The resulting accumulation of fuel, especially in the drier forests, rules out setting a match to the forest, or letting lightning-caused fires burn themselves out. The fear is that even small fires will quickly grow to catastrophic, landscape-scale conflagrations that will destroy all trees and habitat.

The QLG bill prescribes two simultaneous approaches to reduce the threat and create an open, older forest. To protect against catastrophic fires, 40,000 to 60,000 acres per year will be cleared of leaning dead trees, some individual young trees, and some deadfall. Olson says the sites will be cleaned up enough to prevent fire from climbing into the forest crown, but not so much that ground fires can't burn. The QLG bill calls for the work to be done initially in quarter-mile-wide corridors paralleling existing roads. (Olson, who has been working on this plan for several years, says no new roads will be needed on the Lassen.) The corridors will surround watersheds of up to 14,000 acres. The hope is that these corridors will limit the size of fires by providing so-called defensible fuel-break zones, where firefighting will be easier and safer. Once the corridors are completed, the thinning will move into the large blocks of trees defined by the corridors.

George Terhune, a retired pilot who has made himself the group's fire expert, says that the lack of many young trees and leaning dead trees will deprive fire of a path into the forest canopy. Fires will be forced down into the brush or duff on the forest floor, where they should be easier to control.

The removal of deadfall and of small trees and brush will produce some wood for sawmills, but mostly it will produce very small trees whose economic use depends on the development of an ethanol or oriented-strand-board industry. The QLG says it is trying to attract such companies to its area. But at least in the beginning, this work will require subsidies of $100 to $500 an acre, Olson estimates. The second part of the QLG's bill provides trees for sawmills through the very small clear-cuts created by group selection. About 10,000 acres of the 1.6 million managed acres will be logged each year by group selection. The QLG says each acre will be logged every 175 years, but Blumberg estimates that the rotation cycle will be much shorter.

Environmental opponents also argue that it will be hugely expensive, that it will not stop large fires, that it is an enormous experiment being done without adequate science, that it will require new road construction, and that it will double the amount of wood being logged. Finally, opponents say, group selection has often been used to strip forests of their large trees under the guise of opening small plots.

Now that the QLG bill has become law, these issues will be argued out in an EIS required by the bill. John Preschutti, an area resident and former ally of Jackson, Blum, and Yost, says he will participate in the EIS process but he doesn't expect to influence it. "We're at a disadvantage already, because the law will mandate the Quincy Library Group plan," Preschutti says. "So one group has been placed above another. We local people have been sidestepped."

Neil Dion, another local environmentalist, thinks Jackson is making the same mistake the Forest Service did when it drew up its forest plans. "It's a fantasy that they think there is going to be money and resources" and perseverance to do the fire protection. "They will go in and cut the trees and forget about it, and after thirty years we'll be back where we started."

Back to the Past

Which brings the story back to the U.S. Forest Service—a weakened, disoriented agency that used to run the national forests. In recent years the forests have instead been run by judges, by environmentalists working through the Clinton administration, and by the timber industry working through Congress.

Jackson believes that because no one knows how to manage the forests, all sides in Congress hack away at the Forest Service. The Republicans attack science, recreation, and implementation of wildlife protection. The Democrats go after roading and other natural-resource budget items. The result is a shrinking agency that spends much of its time wondering when the next blow will fall.

Jackson may understand why the agency is powerless to set a course on the ground. But he can't resist needling its line officers. "I ask them to tell me which GS-7 (a relatively low job rank) runs this outfit and I will go talk to them."

Kent Connaughton, supervisor on the Lassen National Forest, says Jackson doesn't understand the agency if he is looking for a prime mover. "The Forest Service is akin to an organism," and no one individual can move it. Everyone, Connaughton says, has to want to move in the same direction.

At the moment, that sense of purpose is gone. Until recently, timber provided what Connaughton calls a "clarity" of mission. Now the agency is searching for a replacement: "We're concerned with the biological condition over large landscapes," he says. "I have an idea about where this landscape needs to go—toward the more natural conditions that once existed. But there is still great chaos in how to do it." He says there are also a welter of laws, regulations, court decisions, and microbudgeting from Congress that hinder a forest from moving decisively in any one direction.

Within that big picture, Connaughton says, the Quincy Library Group proposal looks helpful. "When you work through the Congress, you get the validation of the American people. It says that these are a valid set of priorities that we expect you to follow. I appreciate that kind of clarity. It simplifies my life." Long before the QLG bill was passed into law, the Lassen staff was already working on the group's general approach to fire control.

Mark Madrid, the supervisor on the Plumas, is more guarded. He says the bill is a very broad prescription, with the real work still to come. "The meat is going to be in the EIS." But he can see one need: "Our analysis shows it will generate income over the long term. But we will need start-up money."

The Challenge to the Greens

The timber industry and local Sierra Nevada government were driven into Michael Jackson's office by numbers. Through the fat years of the 1980s, the Forest Service sold from 1.5 billion to 1.9 billion board-feet of timber every year, making California the number-two federal timber state in the nation. Then, starting in the Bush administration and continuing under Clinton, timber volume plummeted throughout California. In 1996, only 400 million board-feet were sold in the state—about 25 percent of the total in the 1980s. The decline was mirrored locally on the Lassen and Plumas forests.

If the timber struggle had been a declared war, industry would have sued for peace. But there was no one to surrender to. The timber wars are dispersed, without fixed goals. The industry wants all the forests it can get, and the environmentalists want all the forest protection they can get. In the past, the Forest Service brokered the dispute, but the agency no longer can. Its power has been dispersed to the courts, to the administration, and to Congress. And none of them was available to settle a dispute in an obscure corner of the Sierra Nevada Range.

So the local combatants were forced to deal directly with each other or to remain in perpetual struggle and gridlock. Quincy was not the only place facing that situation. When the timber industry approached the environmentalists in 1992, collaboration and consensus efforts were rare in the West, but they were beginning, and today the West has hundreds of them.

Some of these are true consensus groups, bringing almost everyone to the table. But others divide as much as they unite by destroying old relationships as they strive to create new ones. Rancher Doc Hatfield of Oregon, who formed one of the earliest such groups, says he avoids the word consensus: "These are really collaborative pressure groups. The proposals or strategies that come out of such groups are stronger than the ones coming out of a single-interest group because each participant in a collaborative knows how to bulletproof things against its own side, while a proposal from a special-interest group always has enormous blind spots."

Michael Jackson refers to the way in which diversity leads to strength and divisiveness in different terms: "The environmentalists are mad at me because they think I'm teaching science to the rednecks."

Bonnie Phillips, the head of WaterWatch in Seattle and an observer of collaborative groups, says the Quincy Library Group may be the worst enemy consensus and collaboration ever had because of its decision to go the legislative route. She says, "The other collaborative processes will be seen as incredibly tainted" by what is happening with Quincy. And the timber industry, she predicts, will experience a backlash because of the way it has manipulated the process for its own ends.

The need, she believes, is to put a foundation under collaboration. "The environmental community has to come up with a set of guidelines and principles about when collaboration is appropriate." On the practical level, environmentalists should be trained in hard bargaining. And funding, she says, should be made available to allow environmentalists to participate on an equal basis with industry and government. She also says, "You have to be careful not to play this as the national groups versus the grassroots. National environmental groups don't tell local groups what to do."

Terry Terhaar, now a graduate student at Yale Forestry School, has a different perspective. The former Sierra Club regional vice president for Northern California and Nevada, Terhaar attended Quincy Library Group meetings from March to December 1995, monitoring them as part of her job with the Pacific Rivers Council. She says, "I think there's a lot of fear among environmentalists, especially of the Congress, and for good reason."

But she says another reason for the harsh reaction to the QLG has to do with the structure of the environmental movement. "We all really talked a good line about wanting to help grassroots activity, but what we really wanted was their letter of support for a bill. In my work, I was somewhere in between the national staffs and the state organizations, and I know that a lot of the national staffs don't have time to get involved in gritty details. So if something comes along that they don't like, the easiest thing is for them to blow it out of the water."

Terhaar is especially bothered by environmentalist attacks on QLG, and particularly by those on Linda Blum. Blum, with her husband, Harry Reeves, has been the ultimate unpaid, grassroots environmentalist, whose credentials include service on the board of the Western Ancient Forest Campaign. But that, Terhaar says, has not protected her from attack. Like Bonnie Phillips, Terhaar predicts a backlash, but not against the timber industry. Terhaar says it will be against the campaign that galvanized 140 groups to oppose the QLG. "It's not at all hard to collect a bunch of names—you just say, 'Oh, my god, the sky is falling'—and people sign on. You trust those you think are of like mind." "I predict that when this is all over, some of those grassroots activists are going to sit back and think about what happened. Because Linda Blum is a friend of mine and she's a friend of theirs. And they will say: 'This could happen to me.'"

Montana's Clark Fork: A New Story for a Hardworking River

Donald Snow

For nearly a hundred years, Anaconda smeltermen routinely flushed metallic wastes down western Montana's Clark Fork River. Every spring, when the smelter's settling ponds near Warm Springs were emptied into the massive torrents of the Clark Fork, the river ran red through Missoula. The first "cork" in the river system—the first place where the water slowed to a near halt, allowing sediments to filter out—was a small hydropower dam next to Milltown, just upstream from Missoula. It never occurred to anyone that those sediments, rich in natural arsenic from the copper ores mined at Butte, would find their way into the aquifers around the Milltown Reservoir, but a 1981 discovery of arsenic in a Milltown well proved that they had.

Thanks to an effective effort to locate new water supplies for Milltown, residents there are no longer drinking arsenic-laden tapwater, but the saga of toxicity in the upper Clark Fork basin continues, for Milltown was just the tip of the iceberg, and arsenic was only one among a host of metallic pollutants that line the banks and bed of the upper river.

Within a few years after the Milltown discovery, the Environmental Protection Agency (EPA) designated the entire upper Clark Fork River corridor a federal Superfund site, the largest in the country. In a state renowned for its world-class trout fishing—a state that had just adopted "Naturally Inviting" as its official tourist promotion slogan—the Superfund designation brought dubious distinction.

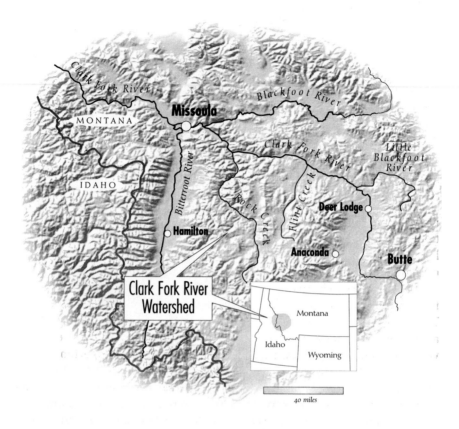

Clark Fork River Watershed

40 miles

This is the new story of the Clark Fork River—not the story of a "pristine" or a "wild" West, but rather the story of a major natural resource, the largest river in Montana, suffering relentless harm from economic actions once considered an unconditional good. But there's more to this river story than just the ongoing ecological tragedy reported in local newspapers and national magazines. In fact, there is a parallel Clark Fork story, largely unreported and unknown, but truly hopeful. It's a story about cooperation, trust, innovation, and willingness to take some risks. Let's start at a new beginning.

Getting Ready

In 1988, the Missoula-based nonprofit Northern Lights Institute (NLI) launched a project aimed at trying to facilitate a negotiated settlement over the Superfund cleanup of the Clark Fork. The battle over jurisdiction and financial responsibility was in full swing, and there appeared to be every possibility that as the fight rumbled from courtroom to courtroom, the needs of the river itself would go neglected.

With start-up funding from the Minnesota-based Northwest Area Foundation, NLI organized a Clark Fork Basin Committee composed of key players in the Superfund debate. The hope was that through a systematic effort in consensus building, the state of Montana, the EPA, and the affected parties in the basin would be able to negotiate an innovative agreement leading to effective cleanup efforts that none of the individual parties could create alone. But it was not to be.

After more than a year of meetings with the Clark Fork Basin Committee, NLI staff and several members of the committee concluded that the project as originally conceived was not going to work. The Superfund fight at that moment appeared utterly deadlocked, with little room left in which to negotiate. Rather than terminate the project, NLI shifted its emphasis toward the least visible component of the original project: a small subcommittee that called itself the Water Allocation Task Force.

The task force suggested refocusing the entire project around the issue of future water allocations, another important, and increasingly tense, struggle. Task force members understood that even as the Superfund issues ground along, the Clark Fork would continue to be a hardworking western river, providing irrigation and domestic water, as well as recreational opportunities to Montana residents and to the thousands of visitors who visit the Treasure State to fish, raft, and enjoy the wildlife, scenery, and community life of the state. Indeed, as task force members well knew, there were now enormous tensions throughout Montana over the clash between traditional uses of the rivers—mostly involving the diversion of water from stream to field—and the increasingly popular new uses demanding that rivers be maintained as living ecosystems.

In the Clark Fork basin, those tensions were becoming especially fierce. For the past twenty years or so, the Montana Department of Fish, Wildlife and Parks (DFWP) had been trying to restore trout habitat in the uppermost reaches of the Clark Fork, where the river's slow meanders through the haylands of the Deer Lodge Valley provide excellent habitat for brown, rainbow, and cutthroat trout. The river's toxicity was one major problem, but perhaps the greater problem was the commonest problem of all rivers in irrigation country: During dry years, rivers and streams are badly depleted, sometimes beneath their capacity to support healthy populations of fish and aquatic insects. In the West, we have a quaint term for this condition: dewatering. It means simply that we're taking nearly all of the water out of our rivers and streams to support irrigated agriculture, lawn and garden watering, and the provision of drinking and industrial-processing water.

By the mid-1980s, DFWP had been studying the problem of depleted flows in the Clark Fork River and had reached the conclusion that the river needed a formal "instream flow"—a special reservation of water to be left in the river in order to maintain and enhance the river's ecological health. This effort on the part of DFWP had reached a crescendo at precisely the moment when NLI's Clark Fork Project came on-line, and was the impetus for converting the project's Superfund focus to a water allocation focus.

The Water Reservation Brouhaha

Under Montana water law, any agency of state government can petition the Montana Board of Natural Resources and Conservation to "reserve" future flows of a major river, stream, or tributary for specified uses. Part of what the reservation process contemplated, when it was enacted in 1973, was the preservation of flows to protect biological, scenic, and recreational resources of the state's relatively pure rivers. But there was more than that.

Montana water savants were looking for innovative ways to protect the state's waters from downstream claims. With the opening of the massive coal-fields on the northern Great Plains, coupled with the aggressive reach of western cities to grab more water to support growth, headwater states felt new challenges to their ability to protect water. Montana's innovative and unique water reservation law provided an attempt to get ahead of the curve.

The state created a means of "reserving" water for specified future uses—some of them attuned to new-fashioned water conservation strategies of the kinds recently favored by federal courts. While the state would declare future reservations for the traditional uses of water—irrigation, municipal, and industrial uses, for example—it would also try to reserve water for fish, wildlife, and recreational uses. Montana was gambling that legal reservations could perhaps replace old-fashioned (and hugely expensive) strategies that western states had long used to cement their claims on water, such as building as many dams and delivery canals as possible. The old battle cry of "use it or lose it" was being challenged by the water reservation idea.

But the architects of water reservations perhaps did not anticipate the virulent reaction their plans would engender among Montana's most powerful water users, the irrigators. Right or wrong, many irrigators see water reservations and the creation of mandated instream flows as a direct attack on existing water rights. By 1989, DFWP stood ready to push for a water reservation in the Clark Fork basin, and the agricultural community stood ready to fight. Since the reservation process is, by definition, a contested case proceeding—pitting one group of scientists and attorneys against another—all the makings were in place for yet another donnybrook over western water.

Enter the NLI Clark Fork project, refashioned according to the ideas of the ad hoc Water Allocation Task Force.

Setting the Table for Consensus

By mid-year 1990, the task force was busily engaged in meetings, under the baton of Project Director Gerald Mueller. With the water reservation petition hanging in the balance, and several task force members standing in clear opposition to each other over that issue, the initial meetings were difficult.

"Tense and stiff," is how Bruce Farling remembers them. Farling, now the state director of Trout Unlimited, was then a staffer with the Missoula-based citizens' group known as the Clark Fork Coalition. Farling recalls the tensions that possessed several task force members who worried that statements they made during the meetings might somehow weaken their organization's position in the upcoming reservation proceedings. "No one trusted anyone, and there was a lot of posturing," Farling told one reporter.

"At that point, we were still locked into the belief that everything about water had to be procedural," says Gerald Mueller. "Too much informality would somehow be regarded as a sign of weakness."

Through regular meetings and increasingly open discussions, members of the task force gradually began to loosen up and build trust, even if many of them still disagreed vehemently about the state's water reservation initiative. By the end of 1990, the group was on relatively easy speaking terms, but it hadn't really done anything, and thus had to face a problem common to many collaborative groups: the possibility that their efforts would amount to little more than talk.

The breakthrough came that winter.

Jim Dinsmore, a rancher from the Flint Creek Valley, stepped forward with a bold suggestion. The upper Clark Fork basin, he said, needed to be closed to the granting of new water rights. His reasoning was simple, but understanding it requires some insight into the workings of western water law.

Under the doctrine of prior appropriation, the holders of the oldest water rights, called "senior rights," receive their water first, while many of the newer rights-holders, especially during dry years, receive no water at all. Still, irrigators who hold senior rights are often threatened by the impacts of new rights-holders coming on-line. During wet years, pretty much everyone with rights gets water, and that often sets up an expectation among even the most "junior" rights-holders that they will continue to be serviced. With no "water cops" to police the basins, senior rights-holders—and their attorneys—are left to their own devices to guard against the intrusions of newcomers.

Dinsmore's notion of a basin closure would help protect senior rights-holders. But instream flow advocates should also see the wisdom of a basin closure, Dinsmore argued, for as long as anyone can file for new water rights—nearly all of them involving the diversion of water from the river and its tributaries—instream flows would also be threatened, and in fact the threats would increase with every new filing.

DFWP and other water reservation enthusiasts at first were alarmed at Dinsmore's suggestion. The reservation they were pursuing was also a water right, and without that right, DFWP possessed no formal standing in many of the basin's key conflicts. But on reflection, Dinsmore's heretical idea of a basin closure began to appeal, for these people understood better than anyone else that the instream flows promised by the water reservation would have been the most junior water right in the basin. And as any irrigator can attest, junior rights in a basin as settled and heavily used as the Clark Fork will never receive a drop of water during dry years—the very years instream flows are most needed. Basin closure would give DFWP at least half of what it wanted: an end to the seemingly endless claims on Clark Fork waters.

With Dinsmore's suggestion, the group now had something upon which it could agree. Closing the basin, a radical concept anywhere in the West, would be good for everyone. That seemingly small step kicked open a new door.

A New Plan Takes Shape

The group began to toy with the idea that the Clark Fork basin needed a full-scale water management plan. A little-known 1967 law called for basin management plans throughout Montana, but early planning efforts had been initiated by state agency personnel and had never garnered much grassroots support. What if a plan were crafted by water users themselves?

But in the Clark Fork basin, with its raging controversies over both water quality and quantity, a grassroots planning effort seemed impossible. As several task force members observed, the reservation proceeding held most of the parties transfixed on the issues surrounding the water reservation. As long as the contested case stood in the balance, efforts to think more broadly about the future of the river and its users would be thwarted. But what if the contested case were postponed? Might it be possible to craft some sort of interim agreement leading to a basinwide management plan while forestalling the reservation initiative?

What grew out of Dinsmore's idea turned out to be a historic agreement among Clark Fork basin water users. They wrote it out and signed it:

• First, DFWP and its supporters among the conservation, hydropower, and sport-fishing communities agreed to suspend the water reservation proceeding for a four-year period, the amount of time the task force believed would be needed to write a citizen-based water management plan.

- Second, the irrigation community, recognizing that some water users might rush to file new claims the minute they learned of the reservation proceedings being suspended, agreed to push for a moratorium on the filing of new water rights during that same four-year period.

These two points formed a classic quid pro quo: Instream flow enthusiasts agreed to postpone the coveted water reservation, but didn't want to give up their place in line among new claims to be filed on the river, perhaps later. The suspension of granting new water rights during the period of postponement answered their request. The agreement as it was eventually crafted and signed by members of the task force had the effect of placing a comprehensive water management plan ahead of the narrower reservation plan.

In January 1991, the task force had its homemade agreement redrafted in the form of a bill and submitted it to the Montana legislature. In April, the measure passed and was signed into law by Governor Stan Stephens. In addition to ratifying the key points of the task force's ad hoc agreement, the new statute called on the director of the Department of Natural Resources and Conservation (DNRC) to appoint a twenty-one-member committee to draft the management plan. What had begun as an effort in "ad hocracy" was now the official business of state government—but without state funding or state control.

Several members of the task force had made it very clear to NLI staff, and to each other, that they were not willing to cooperate in this effort if it became dominated by state government—and in their view, the surest pathway to state agency domination was to have public monies supporting the work. NLI would have to continue to raise private foundation money to underwrite the costs of the effort, or there would be no effort.

In April 1992, the twenty-one members of the new Clark Fork Basin Committee were appointed, with most of the original task force members taking seats in the new committee. DNRC agreed to the task force's original request that NLI be used as the facilitator of the four-year effort to craft a management plan. Thus, NLI continued to staff the committee and underwrite the modest costs of its meetings; DNRC provided the technical staff to perform the research and data acquisition required by the management plan.

For the next year, the steering committee met on a monthly basis and established the mission, tone, and direction of the Upper Clark Fork Water Management Plan. Members of the committee became increasingly aware of the potential for historical significance in their work. This was a real opportunity—perhaps the first of its kind in Montana—for a diverse group of water users to design the destiny of a major river.

The Plot Thickens

During the fair-weather months, Gerald Mueller began organizing field trips so task force members could learn one another's views about the Clark Fork in the presence of the river itself. Those trips provided opportunities for a kind of information exchange that's rare in the narrow, usually indoors, world of policy discussions. Biologists, irrigators, minerals processors, and others were able to inform steering committee members of their organizations' interests in the river. A lot of "native wisdom" came forth—the kind of knowledge gathered up by people on the land, people whose livelihoods depend directly on the river.

During one of those trips—to the beautiful Flint Creek Valley outside of Drummond ("Home of the Famous Bull Shippers")—the committee heard the views of Eugene Manley, undisputed water guru of the valley he has called home for all of his seventy-plus years.

In his soft-spoken but persistent way, Mr. Manley got across his point—counterintuitive to be sure—that simply leaving more water in streams like the little Flint won't necessarily guarantee sustained flows, especially during the normally droughty months of August and September. "The problem with most ideas about these instream flows," he insists, "is that they don't take into account the way return flows work."

Manley points out that today's river basins are heavily altered ecosystems—the products of more than a hundred years of constant change in the hydrological regime. With a sweep of his hand, he indicates the reach of what might be called the "spread riparian." Instead of a concentrated ribbon of water running down the center of Flint Creek's alluvial valley, the "river" is now spread across hundreds of square miles of uplands, through irrigation canals, little dirt-bank ditches, and sprinkler pipes of all descriptions. Eventually, a lot of that water finds its way back to the river, via the river's aquifer, but far downstream from the many points of diversion.

Manley sees the entire Flint Creek Valley as an underground reservoir, a sort of saturated sponge that can be relied upon to squeeze its liquid back into the river system—but only if people cooperate to "irrigate right." "Put on a lot of water early, when you got it," he says, pointing out that if irrigators water heavily during the wet months of May, June, and early July, the stream will reap a bounty of return flows come August. It will take formal hydrologic studies to prove what Mr. Manley believes, but if he's right, there may be a lot more to guaranteeing instream flows than merely shutting down a few irrigation headgates.

Getting the Work Done

Once the planning mission was clearly established and the information gathering was underway, a principal activity left to the committee was to alert the more than three thousand water rights-holders of the upper basin as to the direction of the planning effort. To this end, the new committee conducted a series of public hearings in nine locations and initiated a project newsletter, the *Clark Fork Water News*. More than two thousand copies were circulated free of charge to basin rights-holders.

In addition, the committee realized the need to intensify the focus of the planning effort and to get greater numbers of local water users involved. This was a major undertaking, since the upper Clark Fork basin is composed not only of the main stem of the river, but also of numerous tributaries, some of them with long-established irrigation communities. Through the hearings process, the committee created six watershed subcommittees. They included the Upper Mainstem, the Lower Mainstem, the Big Blackfoot, the Little Blackfoot, Flint Creek, and Rock Creek.

Each subcommittee was asked to identify existing water uses and describe the existing water management system in its area, and recommend actions to resolve water conflicts in its portion of the full watershed. The subcommittees met at night to allow local water users unable to attend the all-day committee meetings to participate in the development of the water management plan. In total, these subcommittees met on thirty-seven occasions and were attended by more than four hundred individuals during a two-year period.

The full committee hammered out its basinwide management plan in a series of twenty-nine meetings that began on October 28, 1991, and ended on December 19, 1994. All committee meetings were noticed through local print, radio, and television media, and all were open to the public. During these meetings, the committee agreed to ground rules to guide its work, explored the interests of its members and the basin, developed a work plan for identifying and addressing basin water issues, and sorted through the recommendations offered by the six subcommittees.

The Results

By January 1995, the steering committee was ready to submit its draft management plan to the Montana legislature. Key provisions of the plan called for a continuation of the ban on new surface-water and groundwater permits, measures to enhance water quality throughout the basin, continuing investigation of water storage opportunities, special attention to water leasing as a means of protecting the basin's fishery, and a ten-year instream flow pilot study to demonstrate the feasibility of using existing senior water rights to provide for instream flows.

The 1995 legislature ratified the plan in total, with the exception of the continued moratorium on groundwater permits. The legislature removed that portion of the water permitting moratorium, feeling that as urban and suburban development proceed in the basin, permits for groundwater should be granted without impediment.

In addition to ratifying the plan, the legislature made a small two-year grant to fund the ongoing work of the committee. The grant came from the Resource Indemnity Trust Fund (mining tax monies), and since by law these grants must go to agencies of government, the grant was given to the Granite County Conservation District, an important actor in the Clark Fork planning process.

Perhaps the most significant provision in the new management plan is its final recommendation—the creation of a ten-year pilot effort in leasing "old" water rights to protect instream flows—for this provision, in a sense, carried the Clark Fork project full-circle. Until the 1995 legislature met, Montana had resolutely resisted the notion of allowing old water rights to be applied to instream flows. Transfers of water use from agriculture (which holds the vast majority of old rights) to virtually any other uses were permissible—except instream flows, the one use that seems to violate the most deeply held precept of water in the West: "use it or lose it." In the eyes of many, instream flows are not a *use*, but a *getaway* of precious water.

But in fully apportioned basins, perhaps the only way to secure instream flows—measurable flows, not just paper water rights—is to have them protected under the rubric of the oldest water rights. And even then, committing water to instream flows is risky business, for any water left in rivers and streams, especially during the driest years, will provide sore temptation to many water users with lesser rights. Once again, the key to the dilemma appears to be cooperation—and cooperation perhaps can best be engendered through the creation of trust and the education exchange that occurs through consensus groups.

The ten-year pilot provision gives succor to those who want instream flows in the river, while at the same time demonstrating agriculture's willingness to try an innovative solution to one of the West's most vexing and long-standing environmental problems—namely, how to keep rivers biologically intact while maintaining a secure economic base for agricultural and other water users. If ranchers can be compensated for the short-term conversion of their senior water rights toward instream flows, everybody wins.

Ratification of the Clark Fork management plan closes an important new chapter in the river's story, but this is clearly a story that never ends. The Superfund effort in the upper basin continues to make news—some good, some bad—while the quieter effort of cooperation in water management captures no headlines. While the Clark Fork project has been a success in the sense that it has carefully built momentum and official recognition for a new, grassroots approach to water management, the project must grow into its next important phase: implementation. And this challenge leads to the final point in our ongoing river story.

And Beyond

The Clark Fork effort, during its first seven years at least, showed that formidable political barriers can be overcome, but now the equally challenging institutional barriers must be addressed. Private foundations and other funders cannot be expected to underwrite the costs of such efforts in perpetuity; they are mostly "seed money" donors, willing to help get things started, but not to provide long-term support. Will legislatures of the West be willing to help finance such efforts? Can small surcharges perhaps be placed on water, such that local watershed groups remain empowered to help resolve disputes and coordinate efforts in cooperation?

These are the kinds of questions that naturally begin to emerge as the Clark Fork project enters the implementation phase. But it's important not to overlook the many questions that have already been answered. Yes, it is possible to overcome the polarization, mistrust, and bitterness that surround water issues in the contentious West. Yes, it is possible to get a conservative legislature to bend on issues many believed to be intractable. Yes, we now begin to believe: It is possible to save a river—even a river with as many problems as Montana's Clark Fork.

The Applegate Partnership: Innovation in Crisis

Cassandra Moseley

Throughout the 1980s, the timber industry, environmentalists, the Reagan and Bush administrations, and federal land management agencies fought over what to do about a small, shy bird—the northern spotted owl. Clear-cutting old-growth forests in Washington, Oregon, and far northern California was rapidly destroying the owl's habitat. By law, the Forest Service and the Bureau of Land Management (BLM) were required to maintain spotted owl populations on federal lands along with all other vertebrate species. But the conservative political appointees heading the agencies were focused on "getting out the cut" rather than species preservation. Environmentalists, wanting to stop clear-cutting, used the spotted owl and the courts to force change. After years of political wrangling, in May 1991, Federal District Court Judge William Dwyer halted all logging on federal lands in the territory of the northern spotted owl and demanded that the government develop a viable owl habitat management plan. In a stinging decision, Judge Dwyer wrote: "The most recent violation of National Forest Management Act exemplifies a deliberate and systematic refusal by the Forest Service and the Fish and Wildlife Service to comply with the laws protecting wildlife."

When the "Dwyer decision" was issued, the division among environmentalists, industry, and land management agencies became a chasm. However, in the Applegate Valley in southwest Oregon, Jack Shipley—a Texas native, former public works director, and longtime environmental activist—realized that the injunction created a rare opportunity. People could no longer fight about timber sales, he reasoned, because there weren't going to be any. Out of this breathing space, then, was born a new coalition of diverse parties seeking better forest management, better known as the Applegate Partnership.

The group's first official meeting was a potluck in October 1992, on Shipley's back deck. The gathering brought together local environmentalists who had been protesting timber sales and herbicide spraying for a decade, and timber company owners who depended on federal sales to keep their mills open. Along with them were federal land managers who had produced those sales and been on the opposite sides of courtrooms from environmentalists but wanted to see change. Instead of introducing themselves as representatives of particular organizations or perspectives, participants spoke about what they loved about the Applegate Valley. As speakers looked out over the valley, they talked about their hopes and regrets. They began to imagine working together to solve some of the valley's environmental and economic problems.

Although this first meeting was key in the founding of the partnership, it was not really the beginning. Jack Shipley laid the groundwork months before by talking to regional and local environmentalists, state and federal employees, county commissioners, community residents, and timber industry folk. He carried a map of southern Oregon from meeting to meeting, explained his ideas for collaboration, and listened to feedback. Among the early supporters was Jim Neal from the Aerial Forestry Management Foundation, a timber group that was trying to reduce logging impacts with helicopters. Neal joined Shipley in talking to activists and leaders from various factions.

One environmentalist recalled meeting with them. "We sat around on sawhorses one afternoon, just kicking around the idea, and they were asking me—from the environmentalist side—whether I though it would fly and that kind of stuff. So I felt included, I mean it was not something that was cooked up outside of what I was doing in the Applegate."

A federal land manager described an early encounter with the Shipley and the partnership idea: "For me, the first time I met Jack or even heard about this project he just walked in, in the summer of '92. I was just in the office. No appointment, he just walked in and introduced himself. He said that he wanted me 'to be aware of some ideas we've been thinking about,' and he started to unfold this map that had all these creases."

On this map of southern Oregon, Shipley had outlined the Applegate watershed. It was the first time that this agency person had ever seen the Applegate watershed as a whole. Shipley went on to explain that he thought it was crazy that so many different agencies and private individuals were managing this watershed without anyone paying attention to what anyone else was doing. He told this manager that some people had been talking about managing the whole watershed as a single unit. "I just looked at him and I

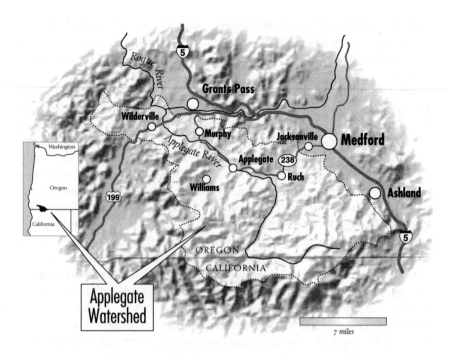

Applegate Watershed

7 miles

thought this guy was insane! And I'm thinking, I've really got to watch this guy, I mean he's talking about doing something that is in my [territory]. All my instincts about power and control and caution came up." At the same time, it was clear to this manager that Shipley had been talking to agency scientists who were interested in ecosystem management. "A lot of what he was saying was making a lot of sense, and yet administratively I was having a lot of trouble." Eventually this manager became a strong and active supporter of the Applegate Partnership.

As Shipley spoke to dozens of individuals, a vision for an experiment began to emerge: Bring people from all walks of life who cared about the Applegate Valley and were willing work to resolve conflict and find common solutions. Another early participant reflected, "Jack really believed that the people that lived and worked in that valley had the capability of solving this problem, better than the agencies alone had."

This early organizing played a key role in the early successes of the Applegate Partnership. When people arrived on Shipley's porch in October 1992, many had already given considerable thought to the partnership idea. After the first meeting, the group developed quickly. At a second meeting, attendees developed criteria for a board: nine members and nine alternates representing a broad cross section of interests including timber, agricultural, environmental, and agency participants. As they were working out the details, one man spoke up with what turned out to be crucial advice, telling the group that "we need to pick people who are looking for solutions and not people who are looking to advocate a particular interest. We need to pick people who are willing to come to the table and be willing to suspend their tightly held ways of doing things." Not their principles, not their deepest values, but their positions and routines. The group listened to that advice and chose several board members interested in building bridges.

At a third meeting, the new board developed a vision statement in a mere hour and a half. Even participants found this speed remarkable: "I've been in groups that have taken months to create a vision, and worked, and teased, and gnawed. But it reflects that Jack and Jim had done so much one-on-one networking—to see where the commonality was and engender and nurture conversation around what those [common] principles were."

Building the Partnership

In the following months, the board met once or sometimes twice a week. Because of the extreme political tension, early meetings were unpublicized, although the group opened meeting participation to all comers after five months. "We had a vision," said one early participant, "though it was a little fragile and embryonic." The group had "to make sure that we were all really solid with the goals and objectives." Participants spent early meetings getting to know one another and developing a sense of what they were trying to accomplish. As one later recalled, "The other thing that they were negotiating was how they were going to play together. What kind of behaviors do we want to embody here at the table that are going to help us meet our goals? People worked really hard to be civil. I mean not polite. But just civil, respectful, engaged, interested."

In these early months, the Applegate Partnership began developing collaborative norms and habits, making explicit agreements about how people would act in meetings. They agreed to "leave their positions at the door and bring only values and interests." To get away from name-calling and blaming, they adopted the slogan "NO THEY," and even created a button of the word THEY with a red slash through it. Group members agreed that people had to come to meetings in a "problem-solving mode." People were not free to just complain—they also had to offer creative solutions. Participants cemented these agreements through practice, by developing habits of collaboration. This was not always easy, for these were emotional times in the forests of the Pacific Northwest. "The rules were pretty wide open. You came as a person.

So personally, [if] you had a lot of frustration and anger about how the world was changing around you and you felt powerless, [that] got played out in the meeting, in the partnership."

Despite constructive early discussion, the group quickly realized that participants could not reach agreements by working in the abstract. They could agree on a few generalized values, but when it came to developing management strategies, sitting in a meeting room got them nowhere. Instead they had to go into the woods. They found that they could agree on cutting this tree and not that one, even when they could not agree on general principles.

One of the first collaborations, called Partnership One, was a jointly planned timber sale. Through numerous field trips and discussions, participants built a project to address recreation interests, provide commercial logs, improve wildlife habitat, and reduce fire hazard. Participants learned about the constraints of each faction—the regulations facing the agencies, the economic and safety issues of timber harvest, and the biological concerns of environmentalists and agency scientists. For the first time, the agencies talked about what would be left behind rather than what would be taken from the woods.

The project pushed the limits of collaboration, and its innovations made many outside the partnership uncomfortable. Even its name aroused concern among environmentalists across the state who feared that partnership groups were going to lead to cooptation. Unfortunately, the ranger district wrote a sloppy environmental assessment and regional environmentalists successfully appealed the sale. By the time the district had revised the environmental assessment and put the timber on the market, the owl injunction had been lifted and timber prices had begun to drop. No one bid on the sale. In retrospect, one agency person reflected that the group probably tried to do too much all at once.

Facing Adversaries

Although the Applegate Partnership became an arena in which participants could work through conflicts and experiment with land management strategies, the group faced some serious challenges. The primary internal ones were what we understand now as ordinary parts of the development of partnership groups—chronic disagreements, time-consuming discussion, and threats of stalemate. The partnership, however, faced additional challenges because early participants were well linked to larger political processes of the Northwest forest crisis. The group became a lightning rod for industry and environmentalist discontent with President Clinton's Northwest Forest Plan.

The main environmental group on the partnership's board, southern Oregon's Headwaters, had been monitoring and protesting timber sales since the late 1970s. It came under pressure from other Northwest environmental organizations opposed to collaboration. Many of the environmental organizations feared that partnerships would lead to weak enforcement of environmental laws. Inside Headwaters, too, board, staff, and volunteers disagreed about the Applegate Partnership's utility. Finally, a fundamental clash between the freewheeling culture of the partnership and Headwaters' highly structured internal democracy strained relations between the two groups. Headwaters wanted to bring Applegate Partnership decisions to its board for approval. Other partnership participants found this unduly bureaucratic, and believed that Headwaters' board members might not appreciate the logic of the partnership's agreements because they had not been involved in the partnership's process. Ultimately, the political and cultural strain became too much; in 1994, Headwaters formally resigned from the Applegate Partnership's board.

Headwaters's well-publicized departure led to editorials questioning the worth of partnership processes. Ironically, however, relatively little changed for the partnership. Chris Bratt, a carpenter and environmentalist, maintained his position on both boards. Jack Shipley, who had been vice president of Headwaters, broke with that group and stayed with the Applegate Partnership. A third Headwater-affiliated partnership board member left and was replaced by an Applegate environmentalist.

As traumatic as Headwaters' departure was for the partnership, it was only a warm-up for more anguished departures later in 1994. The Northwest timber industry was no happier about the Forest Plan than were environmentalists. In an effort to scuttle the Clinton Forest Plan, timber interests decided to sue the Clinton administration for violation of the Federal Advisory Committee Act (FACA), which regulates the role of nongovernmental parties in federal decision making. The court agreed with industry arguments but ruled that the administration's violations were not severe enough to invalidate the Forest Plan. Instead, the court told the administration that it had to change its operations. Reflecting an excess of caution, the administration directed the partnership's federal agency board members to resign, and suggested that they could no longer even meet regularly with the partnership. For partnership participants who believed that they could succeed only with federal involvement, this ruling was an enormous blow. Said one: "FACA was huge. FACA just about killed the partnership. In fact, we thought it was going to."

Over the next few months, civilian participants and federal employees searched for ways to work together. Eventually, the White House eased its interpretation of FACA's restrictions and even encouraged agency-citizen collaborations as long as decision making over federal lands continued to rest clearly in the hands of federal employees.

Somewhat surprisingly, the crises of 1994 did not kill the Applegate Partnership (although there were many such rumors in Washington, D.C., that it had). Rather, they made the group stronger: The federal agencies did not disappear, but rather found ways to work with the partnership and the rest of the Applegate community more effectively. As one participant remarked, they "had to search for new structures and actually communicate with the feds. The partnership had to get much more precise and specific about what kind of help and assistance in partnering they wanted to do with the feds; and the feds had to get really clear about what they could do and could not do, how they could participate. And that ended up being, in some ways, a great strength."

Changing Federal Agencies

Even with the FACA restrictions, the partnership and agencies found considerable opportunities for collaboration and innovation. In effect, the partnership helped the agencies realize that they would "pay now" with up-front participation and innovation or "pay later" with time-consuming appeals, court cases, and civil disobedience.

The Ashland Resource Area of the Medford District BLM responded to the challenges of the Applegate Partnership by transforming its timber sale planning process. To address the partnership's concerns with piecemeal planning, the resource area moved to a system of subbasin planning with an eye toward "treatment across the landscape." Area managers began to involve community residents early and often. The resource area's first project under the new system was in the Thompson Creek drainage, where residents and agency planners were concerned about the hazards created by decades of fire suppression. Planners took community residents, timber industry representatives, environmentalists, and scientists to the site to talk through ideas and get feedback. After developing a preliminary plan, the resource area undertook a test harvest so that everyone could visualize the larger harvest. More field trips, meetings, and modifications followed before the project went out to bid.

Procedurally, the project was a great success. A wide variety of people voiced their desires and concerns early in the process. The agency created a sale that environmentalists did not want to appeal and timber companies were willing to buy. Even in its flawed outcomes, resource area managers learned valuable lessons for future sales. (Among the complaints: Too many trees were harvested from ridgetops, creating new paths for off-road vehicles and weed invasions; and not enough trees were harvested downslope, where fuel loads remained too high.) In the next planning area, the resource shifted prescriptions to address problems discovered in Thompson Creek.

One of the greatest challenges, however, was figuring out how to cover the non-revenue-producing parts of the projects, especially thinning the brush fields. These areas would once have burned every few years but were now so overgrown that they had even lost their value as wildlife habitat. Traditional restoration funds were not available, as these dollars were tied to timber sales. Initially, the resource area used special funds associated with the Northwest Forest Plan, but the managers knew that this funding was not going to be around for long. With Applegate Partnership backing and with support from the district manager and state office, the resource area manager went to the Washington, D.C., BLM office and got permission to use restoration funds for broader purposes.

Whereas the BLM's Ashland Resource Area remade its management system to be more participatory and ecosystem-based, the Forest Service's Applegate Ranger District became an arena for a variety of innovative experiments. For example, the ranger district, the partnership, and the Applegate River Watershed Council negotiated a contract in which the watershed council would write an environmental assessment for the management in the Carberry Creek drainage, a high-elevation subwatershed. As part of the planning process, the watershed council undertook a landscape design that it hoped would help tie timber sale planning to long-term ecological objectives. Although this project proved challenging, a management group within the district later used lessons from the Carberry project to build a multiyear landscape design and implementation effort in the much larger Little Applegate drainage.

In another example of this experimental approach to national forest management, there has been serious discussion about abandoning timber targets, which agency personnel and community residents have often blamed for unecological management. Although it has not yet been implemented, the ranger district and the Applegate Partnership have talked about undertaking a two-year experiment using ecological management goals rather than timber targets to drive management.

Although the ranger district has often been a seat of innovation, the structural divisions inside the Forest Services meant that the ranger district sometimes also continued business as usual, including undertaking controversial timber sales without much advance notice. One such controversy erupted when the district ranger proposed a timber harvest in an unroaded area known as Collins Mountain. Early in the Applegate Partnership's history, federal and civilian participants agreed that they would avoid harvest in these areas because they were areas of "nonagreement." Several years later, however, the district ranger, Mary Smeltzer, had to meet her timber targets and hoped that the harvest in the Collins Mountain area would help. Environmentalists objected vehemently and asked Smeltzer to live up to this earlier agreement. Frustrated but willing to negotiate, Smeltzer challenged environmentalists to identify other areas in the district where she could harvest without uproar. Headwaters and the partnership met her challenge and eventually brokered a deal where Smeltzer would not immediately harvest Collins Mountain, but would revisit the region in a few years with an eye toward harvest.

Brokering deals such as these was a frequent part of the Applegate Partnership's work. When controversies erupted, the partnership encouraged negotiation. Often the first step was encouraging the particular agency to pause and consider more fully residents' concerns. Community support of particular projects often emerged when the partnership and agency personnel could find concrete ways to address larger issues in different arenas.

In contrast to the example set by the Ashland Resource Area and the Applegate Ranger District, the BLM's Grants Pass Resource Area responded to the Applegate Partnership by retrenching. Initially, the area manager had been involved actively in the partnership, but when she was promoted to Portland, her replacement, Bob Korfage, pursued more traditional management approaches. He arrived after the spotted owl injunction was lifted, and found it difficult to lay the groundwork for reform. In his rush to meet increasing timber quotas, Korfage adopted a "pay later" approach by minimizing public involvement and asking for feedback when planning was already well underway. This mobilized local environmental activists: By 1998, all but one of the area's timber sales were under administrative or legal appeal or had been delayed because of community uproar.

Growing Pains, Looking Ahead

The Applegate Partnership came together in a period of enormous stress and crisis. The old politics had broken apart and created an opportunity for change. A handful of energetic leaders in the Applegate Valley came together, spending huge amounts of time and energy to remake land management. Nearly a decade after the partnership's founding, however, many of the original participants have moved on. Many timber industry people left after the injunction was lifted or their businesses failed. Other founders moved onto jobs outside the valley or simply retired. Although Jack Shipley and several of the early valley residents remain, one of the key questions facing the partnership today is how long it can or should remain together. For some of the early founders, including Shipley, the partnership wasn't supposed to become bureaucratized; Shipley saw it as a group that existed only as long as it had work to do and the energy to do it.

The period of rapid innovation and change in the federal land management agencies has passed. The partnership continues to play an important role as a negotiator, a locus of public involvement, and advocate for continued innovation. Moreover, the partnership has taken on the role of watershed council, implementing half a million dollars of private land restoration funds each year, and has been instrumental in building linkages between public and private land management efforts. If the partnership continues, it must shift from a group fueled by crisis to one that can maintain itself for the long term, while keeping the innovation and creativity that marked the early years.

Malpai Borderlands: The Searchers for Common Ground

Kelly Cash

The New National Gallery in Berlin is showing John Ford's 1956 western, *The Searchers,* on a huge projection screen. Visible to street traffic day and night, the movie, shown at one frame every fifteen minutes, will take over five years to run from beginning to end. *The Searchers* is above all a quest film. It follows the story of Ethan Edwards (John Wayne), an ex-Confederate soldier who returns three years after the war to the family ranch in Texas. Within days of his arrival, a Comanche massacre kills most of his relatives and turns the reunion into a funeral. His niece, nine-year-old "Little Debbie," has been taken captive. For five years Ethan searches for her. When he finds her at last, she is a young woman (Natalie Wood!) and the wife of Ethan's nemesis, the Comanche Chief Scar. The climax of the movie is about overcoming hatred and rediscovering humanity.

For those of you whose cinema history is rusty, or who have never been able to get past the film's disturbing cultural politics, *The Searchers* is the apogee of the John Ford western, meaning that John Wayne and the landscape are equal stars of the show. Filmed in Colorado, Alberta, Canada, and Monument Valley, Arizona, the story derives much of its power from its setting.

As everyone knows, these VistaVision settings are becoming rare and astronomically valuable. Ranchers in Jackson Hole, Wyoming, are offered $100,000 an acre today for ranchland valued at $10,000 a decade ago. Once acquired, the Ted Turners and dot.com millionaires often put telephone and electricity wires underground to preserve the viewshed. This impulse also has a more populist streak, as millions stream into national parks to experience epic landscapes that seem unchanged by human hands.

The story I want to tell is of a handful of men and women—ranchers whose lives are played out against just such a place, and whose quest to remain there will require an odyssey as grueling as Ethan Edwards's, but with a twist. Edwards became a symbol of individual frailty in the face of an isolating "wilderness." But for the ranchers of the Malpai Borderlands, it is the wilderness that has become frail, and the survival of all they hold dear will depend not on rugged individualism, but on their newfound ability to forge and sustain literally thousands of relationships.

Ranchers concerned about the future of cattle ranching banded together in the early 1990s to form the Malpai Borderlands Group, a nonprofit organization. Through a series of front-porch meetings and serendipitous relationships, these ranchers have created one of the largest experiments in what scientists call "ecosystem management" in America today. The group's mission statement reads: "Our goal is to restore and maintain the natural processes that create and protect a healthy, unfragmented landscape to support a diverse, flourishing community of human, plant, and animal life in our Borderlands Region. Together we will accomplish this by working to encourage profitable ranching and other traditional livelihoods, which will sustain the open space nature of our land for generations to come."

The group has begun to work *mano a mano* with a sea of cooperators, including state fish and game agencies, the U.S. Department of the Interior, the Department of Agriculture, universities, and several nonprofits such as the Animas Foundation, The Nature Conservancy, and Environmental Defense. Within a few years of its founding, the effort had received the endorsement from the highest scientific and environmental circles, including the Society for Conservation Biology's Distinguished Achievement Award. Articles about the group appeared on the cover of *Audubon* magazine, *The Nature Conservancy* magazine, the *New York Times* science section, and *Smithsonian* magazine. Not bad for a place that got electricity just ten years ago.

After the Martians Landed

The group's project area is shared by southwest New Mexico and southeast Arizona. It is roughly triangular in shape, the bottom bordering 60 miles of the international boundary with Mexico, the west ending at the borders of the Gray Ranch, the north peaking a little past Rodeo, New Mexico, and the east bounded by Highway 80 out of Douglas, Arizona. Tucson is only a two-hour

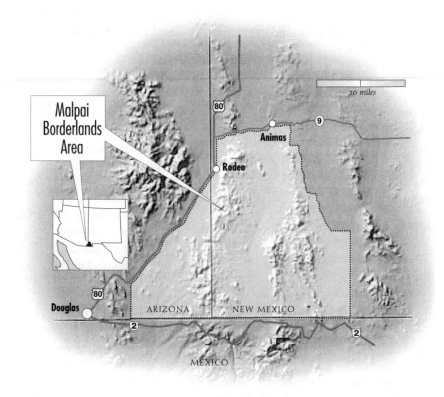

drive away. One of the best ways to encounter the Malpai Borderlands is to drive through Douglas, perhaps stopping for a beer at the Saddle & Spur Bar in the Gadsden Hotel, and then heading east onto Geronimo Trail Road. Once you skip into the hills west of town, you are greeted with an Alaska-like encounter with vastness so complete that giddy euphoria seems the most appropriate response.

It is here, in this roughly one-million-acre Malpai Project Area, that the Sierra Madre ends its 1,200-mile journey that begins in southern Mexico. It collides with the southernmost point of the Rocky Mountain range, whose northernmost tip lies 1,400 miles to the north. The Malpai region includes flora and fauna from no fewer than seven biomes, including the Mojave, the Sonoran, the Chihuahan, and the Great Plains. One glance can take in ocotillos, pine forests, and oak trees. A century ago there were healthy populations of Mexican wolves and grizzlies, with thick-billed parrots only a valley away. Today there are six federally endangered species and more than sixty considered threatened regionally. More species of rodents occur here than in several states combined, and there are more lizard species here than in any other location in the United States. In 1996, the Malpai Project Area was even visited by a Mexican jaguar. It is the kind of landscape that makes environmentalists swoon.

Which is exactly what The Nature Conservancy did in 1989 when the absentee owner of the Gray Ranch gave the Conservancy two weeks to buy the property, or watch it fall to the chopping block. The response from Conservancy supporters, from the pennies of schoolchildren to five-digit contributions, was incredible. The Conservancy was able to pay $18 million cash on the barrelhead, at the time the largest conservation purchase in history. The Gray Ranch was "saved." The response from the environmental community was unmitigated jubilation.

But in local communities, the response was quite different. "You might as well have told us the Martians had landed," neighboring rancher and charter Malpai founder Bill McDonald later said. For the Conservancy, "Plan A" was to sell the Gray Ranch at cost to the U.S. Fish & Wildlife Service, for designation as a national wildlife refuge. After listening to local objections to this idea, then Secretary of Interior Manual Lujan made it clear that a "Plan B" was needed. Enter John Cook—a "bookish Easterner," in the words of McDonald—who was sent by the Conservancy in 1992 to "fix" the Gray Ranch. Equal parts entrepreneur, rainmaker, poet, and fifteen-year Conservancy veteran, Cook entertained a few serious inquiries from private buyers, including Ted Turner and Jane Fonda. But in the end it was a call from a local rancher, albeit an extremely unusual local rancher, that determined the future of the Gray Ranch.

Drum Hadley is a rancher and poet who has lived in the Malpai Borderlands for thirty years. His children, including his son Seth, were born there. Among literary and conservation circles he is known as a kind of vaquero Gary Snyder. While the Conservancy was looking for a conservation buyer for the ranch, one of Hadley's neighbors—acting as an emissary for local ranchers who knew the Hadleys had means beyond their ranching operations—asked if Hadley could buy the Gray Ranch. It seemed an impossible proposition, but then Seth Hadley proposed that they could use their inheritance to create a foundation (to be called The Animas Foundation) to "encourage the practice of land ethics to preserve, heal, restore and sustain wildlands and waters, their inhabitants and cultures." The final decision fell to Drum Hadley's mother. When Drum explained what he and Seth wanted to do, his mother replied, "Do you have a pen that writes?"

Kitchens That Changed My Life

Years later, John Cook would comment in his *Working Notes: A Hybrid Approach to Conservation*: "In the two years of work in structuring the sale of the Gray from TNC to the Animas Foundation, the Hadley Family and the

TNC team worked hard enough and long enough together to establish common ground values together. This is another way of saying we built trust. That gave Drum Hadley the confidence to open the door and invite TNC through to meet the neighbors. On the other side of that door were individuals with whom he had twenty years of neighboring. They took him seriously when he suggested we might have something we could bring to their table."

John Cook's father had been an Episcopal priest at Harvard in the 1960s, and John himself has a gentle yet astute way of holding a group of people together. It was John's ability to supply the know-how to put a nonprofit group together that helped transform the group's front-porch discussions into strategy, capital, and action.

Trawling for a helping hand, John invited me out to tour the project when I was four months' pregnant. This was in the summer of 1994, back when the Sagebrush Rebellion was red-hot, and my dealings with ranchers back at the communications department of the Conservancy's national office was termed "Opposition Management." Since the Conservancy had sold the Gray Ranch, and didn't own an acre of land in that part of the world, I didn't understand what we were doing there. I had worked for the Conservancy for fifteen years. Where was the future preserve? Where was the green oak leaf? John looked at me. "Your job," he said, "is to shut up and sit in the corner." Over the next three days, sitting in the corner turned out to be sitting in four kitchens that changed my life: the McDonalds', the Magoffins', the Glenns', and the Millers'.

For ninety years the McDonalds have lived on Sycamore Canyon Ranch. Bill McDonald is a fifth-generation rancher whose father was an English professor. A large, handsome man of Mormon descent, he learned his ranching skills from his grandfather. Like many ranchers, he has an uncanny ability to express his thoughts plainly and sensibly, but then he will throw in a word like "psyche." In 1998, he became the first rancher to receive a MacArthur genius award for his statesman-like ability in creating what he calls "the radical center." At lunch, he and his wife, Mary, equally talented and a microbiologist by training who now oversees the group's complicated finances, talked about why the community decided to form an official organization. Bill said, "We'd gotten awfully good at knowing what we were against, and decided it was time to figure out what we were for."

In reaching out to its critics, the group discovered that one primary thing that everyone was "for," was fire. To understand this, remember that this region is influenced by the Great Plains biome, where prairie fires are commonplace. In fact, the movie *Oklahoma* was filmed just two valleys to the west of the project area.

We've all heard it said that the West is a checkerboard of public and private ownership, providing the image of neat, interchangeable squares. The reality often looks much more like some kind of optical puzzle: a large square of Forest Service land here, a thin rectangle of state land there, a zipper of private riparian zone cutting back to Forest Service or BLM land at the higher elevations, the valleys mainly private. In the Malpai, the different squares add up roughly to 53 percent of the land in private ownership, 23 percent in Arizona and New Mexico state ownership, 17 percent owned by the Coronado National Forest, and 7 percent by the Bureau of Land Management. The gridlock produced by this ownership pattern had resulted in eighty years of fire suppression. This, combined with historic overgrazing and a documented climate shift, fed an invasion of woody shrubs into an area where wild hay operations had flourished at the turn of the century. It wasn't until the Malpai Borderlands Group was able to get everyone in the same room together that a nearly 850,000-acre "fire map" for the region could be developed, so that wildfires can burn on private property with the landowners' consent. Larry Allen of the Coronado National Forest estimates that the plan easily saves taxpayers more than $1 million per year.

Next, the Malpai Group and their new cooperators tackled what had previously been unthinkable: a prescribed burn, requiring the blessings of four private landowners, five government agencies in Arizona and New Mexico, international coordination with Mexico and adherence with the Wilderness Study Area, National Environmental Policy Act, National Antiquities Act regulations. Planning and execution of the "Maverick Burn" were accomplished in eight months.

Saving All the Parts

If the success of a collaborative approach to fire was the first head-turner of the Malpai Project, water as well as the power of the single ranching family was the second. This is a landscape where 8 inches of rain is a good year, where one million acres of mountains, some over 8,000 feet high, do not produce a single perennial creek. Down the road from the McDonalds live the Magoffins, a family of ranchers who have worked for years with biologists and herpetologists to preserve one of the world's last populations of the threatened Chiricahua leopard frog, which occur in stock ponds on their land. In the early 1990s, when successive droughts threatened these ponds, the Magoffins hauled more than a thousand gallons a week to keep them from drying up. From the Magoffin kitchen, you can see a taxidermy mount of a snarling javelina head with a vintage hat perched saucily on its head. "Pig with PMS," says Anna. But when I ask why they went to so much trouble for a frog, she replies, "The frogs had become our friends. We couldn't watch them die."

Working with the Malpai Group, the Magoffins secured funding from Arizona Game & Fish and Malpai supporters to drill a well for permanent water that would support both the frogs and the Magoffins' cattle. I remember a particular "Malpai Moment" when I watched Matt Magoffin talking to herpetologist Phil Rosen about fencing the cows out of half of a pond to make sure that the cows didn't affect the frogs. Matt then went on to describe watching the frogs disperse during a monsoon, hopping a quarter of a mile in thirty minutes. Phil said, "that's the kind of thing I would never see because I don't live here."

Historian Patricia Limerick has written, "It is difficult for most people to realize that a particular moment in 1869 was once the cutting edge of the present—-the edge beyond which the future could only be glimpsed by guessing." Rancher Warner Glenn seems to defy this. John Cook remembers talking with Warner one day "about what a time of upheaval the Sixties were, how tough it was on so many families, and Warner agreed. Later on, we realized that I was talking about the 1960s, and he was talking about the 1860s." Warner's partner, Wendy, has a sense of the past and future that equals his. One day while checking water at a pump, Wendy talked excitedly about the native grasslands restoration that she and Warner had begun with the Malpai Group's help. She spoke about the high concentrations of hawks and grasshoppers she had been watching. Then she became uncharacteristically wistful for a moment, and seemed to be talking about more than just the restoration on the Malpai Ranch. "Sometimes I think about how much time I have left and it makes me sad. Not because I have to die, but because I won't get to see everything that's going to happen."

It was at the Glenn's Malpai Ranch that the Malpai Borderlands Group came into being, and the first office was housed in one of their bedrooms. Much of the dialogue that drives the group still takes place there. Their kitchen is a meeting ground of exquisite food, long conversations, small dogs, children, ranchers, hunters, environmentalists, public agency staff, and the occasional rattlesnake. Kitchens, porches, dogs—family ranches integrate life and business in a way that is foreign to most of us. This also helps explain ranchers' distrust of "officialdom," which seems far removed from their world. At the same time, this is one of the reasons why collaborative conservation is so effective, creating a forum for ranchers and outside mediators such as The Nature Conservancy and Environmental Defense, to help negotiate the labyrinth of environmental laws and agency "officialdom." This is especially important in areas like the Malpai Borderlands, where much of the land is managed by the Coronado National Forest, which has more endangered species than any other national forest and, not surprisingly, is also one of the most heavily sued national forests in the United States.

Individual endangered species have the power of the law behind them; ecosystems don't. The core grasslands restoration effort that helped catalyze the group was the reintroduction of fire, but the group quickly collided with the Endangered Species Act, which can unwittingly deter the conservation of species it has set out to save. Before my interaction with the Malpai, my faith in the Endangered Species Act was complete. Now I watch lawsuits launched by well-intentioned environmentalists that, if won, would jeopardize the populations of the very species they are trying to save. For example, many green groups list the Chiricahua leopard frog as a victim of cattle ranching. Yet most herpetologists recognize that isolated, bullfrog- and bass-free stock ponds are this leopard frogs' best hope of survival. Michael Bean of Environmental Defense, a longtime expert on the Endangered Species Act, captures the tragic folly of these lawsuits: "If you measure your victories by the lawsuits you've won, you may be making the same mistake that the U.S. Army made using Vietcong body counts as proof that we were winning the Vietnam War."

The reality is that the huge areas needed for "cores, corridors, and carnivores" require working with multiple landowners, especially stabilizing land tenure when good or promising managers are already there. Information and energy on behalf of a species can move in informal yet powerful ways. For example, in the case of the Chiricahua leopard frog, Anna Magoffin's brother, Hans, happened to teach biology at Douglas High School. Hans enlisted the help of his students to rear young frogs and then build a frog pond for them on the school's campus. Anna's and her brother Hans's hope was that the frogs could then be transplanted to other stock ponds, but regulatory concerns about the species have preempted the effort. Similar problems have plagued the Malpai Group's work with fire. How will fire impact the ridgenosed rattlesnake? The agave that the long-nosed bat depends upon? Is there suitable Mexican spotted owl habitat in the area? There seems to be an endless array of agencies and lawyers that need clear answers to these questions, and it doesn't matter that these species have evolved in habitats that have burned for millennia.

Out of the Kitchen, but into the Frying Pan?

Many of my environmental brethren have berated me for "mission drift," for being seduced by the cowboy myth as the result of my work with the Malpai. You can enumerate all the reasons that working with ranchers is a sound environmental strategy to achieve conservation goals (and vice versa for ranchers), but there is a sly bigotry at work that prevents many people from believing that it's possible. One colleague asked, "You can kick the dirt with the ranchers, but what do you have when it's over?" For the growing number of environmentalists who believe the answer is "more than we could ever get on our own," the results from the Malpai to date are impressive ammunition.

The Conservancy began by buying a 324,000-acre property to protect a "nature preserve." But instead of entering into a hybrid relationship, and throwing our might into creating a "working wilderness," five years later we now see one million acres in cooperative conservation ownership, including several additional ranches in Mexico. Conservation easements are now held on 30,000 acres of private land affecting 60,000 acres of adjoining and inter-mingled public lands. The Animas Foundation has reintroduced black-tailed prairie dogs on the Gray Ranch, twenty-four bighorn sheep have been released on Bill Miller Jr.'s ranch, and there are two hundred permanent monitoring sites established to gain information on how to promote general ecosystem health. The group's Jaguar Fund is helping to support research and inventory of jaguar habitat in northern Mexico and in its own project area. In fact, using science as a "community-building tool" has proven so effective that researchers supported by the National Science Foundation and the African Wildlife Foundation are studying Malpai lessons for possible export to Kenya.

Seeing the conservation power that motivated private-sector ranchers can ante up helps me understand why Bill McDonald looked so puzzled the first time he saw a scorecard sheet from the New Mexico Heritage program, detailing species on "protected" versus "unprotected" land. He asked, "Why is everything on private land listed as 'unprotected'?"

Science is usually thought of as the purview of the cap-and-gown or Teva sandals crowd. But the Malpai Ranch hat rack tells a different story: White Stetsons coexist with baseball hats advertising obscure scientific groups, such as the American Society of Mammalogists. The heads that these hats belong to have stepped into what John Cook has called "'the hybrid': a partnership of unexpected and surprising stakeholders determined to move a common agenda ahead no matter what it takes."

And as the Malpai ranchers and supporters are finding out, it takes a lot. Even the most creative and useful ideas—such as Drum Hadley's grassbank concept, in which grass on one ranch can be exchanged for conservation action of equal value on another ranch—take a lot of time, legal help, and capital. The process of protecting their "traditional lifestyles" has done much, ironically, to almost wreck them. The Malpai Group has had to adapt to an almost urban, incessant "busy-ness"—phones, faxes, emails, direct appeals, newsletters, meetings, memos, visitors—that their ancestors certainly never envisioned. And their enemies on both sides have become more voluble and vitriolic as their successes mount.

Yet despite these trials, the group continues its search. In fact, members often point to the group's formation as giving them a new opportunity to engage in old-fashioned "neighboring." It is their belief that the strength of this same neighboring, which secured a school bus route and electricity, will somehow see them through to their dream of staying on the land. For those of us at ringside, so cognizant of the market forces transforming so many other places in the West, their story often seems much more of a cliffhanger. But the ranchers of the Malpai Borderlands Group seem to have no doubt that they will succeed. They are, as Ethan Edwards might say, "as sure as the turning of the Earth."

Colorado's Yampa Valley: Planning for Open Space

Florence Williams

On his ranch near Oak Creek, Colorado, Dean Rossi leads a small group along the banks of the Yampa River. They are not talking about the hay crop, cows, or the slumped cattle market, all on the forefront of Rossi's mind. Instead, they are talking about stream velocity, erosion, and the regeneration of cottonwoods. Rossi, forty-nine, listens intently as Holly Richter, a scientist with The Nature Conservancy, describes the river system.

"I'd like to see more young cottonwoods and willows so that over time this is a sustainable plant community," says Richter. "What would you like to see?"

Rossi fingers his worn red cap. "I have no problem with more willows here as long as they don't take over my meadows," he begins. "I'd like more cottonwoods here, that's better shade for the cows."

Down the river a quarter mile, Rossi points to a curving bank that has eroded steeply, taking with it some of his prime pasture. "This is my weakest link here," he says.

Richter nods. "Maybe we can put in some boulders in the stream here and slow the flow, get some willows rooted in these banks. From my perspective ecologically, these vertical banks are a problem."

Much to their mutual relief, Richter and Rossi, who head the local cattlemen's association, have found some common ground. This meeting was not serendipitous, but the culmination of nearly a year of careful negotiation, fund-raising, and hard work. In May 1995, Rossi and his brother Jim had part-sold, part-donated a conservation easement to the local Yampa Valley Land Trust that will forever protect 4 miles of the Yampa and its fragile riverside forest. In doing so, the Rossis set off a string of accomplishments: This was the first purchased conservation easement on agricultural land in the state of Colorado, the first ranch easement funded by state lottery money and by the local county, and one of the few private ranches in the state to jointly manage its riparian habitat with a nonprofit conservation group (The Nature Conservancy recently drafted a management plan).

"For us, this agreement really made sense," says Rossi, whose family has farmed and ranched in Routt County for three generations. In exchange for agreeing never to develop the river corridor or sell the property to developers, the Rossis received a generous pile of cash, a substantial tax write-off, and the knowledge that their beloved ranch will never buckle to condos and asphalt. "The money will enable us to buy more lease land, which we need," says Rossi. "And I'd much rather see open-land ranching here than a bunch of houses, which would have been the alternative."

Such win-win achievements are becoming familiar in this part of northwest Colorado surrounding the town of Steamboat Springs. By 1997, more than 10,000 acres in Routt County had been set aside as perpetual ranchland through conservation easements. In November 1996, Routt County became the first in the intermountain West to approve a tax-increase initiative specifically for purchasing such easements. Valleywide partnerships have secured more than $15 million in state and federal monies for such projects as rural telecommunications, transportation planning, parks and trails, wildlife, and open space.

Thanks to an infusion of funds, a can-do community of activists and citizens, and a highly responsive local government, Routt County has become a model community on many fronts. At the start of the twenty-first century, it sits confidently poised on the cutting edge of conservation, diverse economic health, and effective, productive coalition building.

It wasn't always this way. Just half a dozen years ago Routt County citizens were torn apart by a proposal to develop a huge new ski resort 6 miles south of the existing Steamboat ski area. The pro-growth and no-growth camps resorted to a less-than-lofty battle of bumper stickers. "Stop the Brutal Marketing of Steamboat," read one. The local chamber of commerce responded by issuing to hundreds of locals the sticker, "Steamboat Welcomes You." Opponents could hardly sit in the same room together, or at the same traffic light. Visitors to the area, if they noticed the goings-on of their hosts, simply scratched their heads and planned to go to Vail next time.

What changed?

"We got educated," simply states Arianthe Stettner, a city council member and local activist who fought the Catamount ski area proposal. "We eventu-

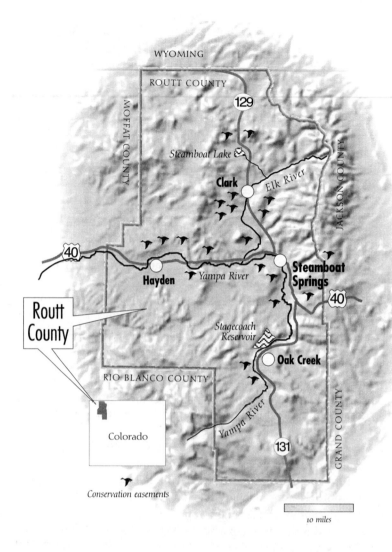

WYOMING

ROUTT COUNTY

MOFFAT COUNTY

JACKSON COUNTY

129

Steamboat Lake

Clark   Elk River

40

Hayden   Yampa River   Steamboat Springs   40

Routt County

Stagecoach Reservoir

Oak Creek

RIO BLANCO COUNTY

GRAND COUNTY

Colorado   Yampa River   131

Conservation easements

10 miles

ally learned about public process through the Catamount debate, and we learned that being shrill doesn't work. We now know we have to be constructive, offer sound alternatives, and be nonconfrontational." Stettner got literally educated as well, earning a long-distance business degree with a specialty in "sustainable development," which she defines as a healthy local economy that still protects critical resources. She returned home to chair the Vision 2020 project, a multiyear planning effort by a diverse group of citizens to define what qualities of life they wanted protected or enhanced in the Yampa Valley, and to make recommendations for achieving them.

As in many resort communities around the West, residents here perceived spiraling growth, and with it such concerns as air quality, road congestion, stressed schools and services, escalating home prices, and rising taxes. During the early nineties, Steamboat's population grew 6 percent a year (it now stands at 8,000, with another 8,000 in the county); between 1993 and 1994, county building permits increased 18 percent, and home values during that time increased 36 percent. Because of Colorado Senate Bill 35, which exempts from review any subdivisions of greater than 35 acres, parts of rural Routt County were beginning to resemble pepperoni pizza with too much meat, a house on nearly every ridge.

"Vision 2020 was about building consensus," Stettner says, "and it worked. People bought into it." Among the final report's conclusions, issued in 1994, were these: Maintain a sense of community, preserve open space, create a sustainable economy, provide sensible transportation, and engage in comprehensive planning.

Those were about the same conclusions being reached in other corners of the community as well. A local group called Environment 2000 began sponsoring annual "nonconfrontational" conferences to work through such topics as "Where do we grow from here?" C. J. Mucklow, Routt County's agricultural extension agent, also saw the need for building bridges between camps. He organized an "eco-ag" committee with representatives from environmental groups and the ranching community to sit around a table every few weeks and just talk. "It was a little tense at first," he recalls, citing an early discussion about vegetarianism. "But those individuals ended up really respecting each other. There developed a real synergy."

Out of such coalitions (and their recommendations) sprang more effective talkfests and government-sponsored public-process endeavors: a community comprehensive plan (endorsed by both the city and the county), an open-lands plan, an intermodal transportation plan, an affordable housing plan and an economic development plan. There was even a local joke that people were "blinded by vision" as well as regular groans of "Oh no, not another visioning process." As city council president Kevin Bennett puts it, "We're big on meetings."

But documents and policy statements alone do not effect change. In small communities like those in Routt County, there has to be political will, and political will soon materialized. "In the last few years, the city and county governments have been incredibly helpful and incredibly supportive," says Susan Dorsey Otis, director of the six-year-old Yampa Valley Land Trust (now housed in an office donated by the city). "There's nothing like having elected officials who support the public and support citizen initiatives. It's very empowering, and it keeps people involved."

Citizens, for their part, took up the planning ball and ran with it. In the upper Elk River Valley north of Steamboat, six large landowners convened in 1993 to talk about how to protect ranches from encroaching development but still avoid regulatory mechanisms. The farsighted group ended up donating conservation easements totaling 2,800 acres to the American Farmland Trust, a national agricultural land trust. In so doing, they set an example for their neighbors and engaged local government in their commitment to land preservation. "We let the county commissioners know what we were doing and why," says Jay Fetcher, who, with his eighty-five-year-old father, runs a cow-calf operation. "We wanted to get their buy-in so that if someone proposed a mobile home park on irrigated meadow, the commissioners would know that's not what we want." Furthermore, when Fetcher later helped organize local opposition to a proposed ski-touring development, people listened, including the U.S. Forest Service, who had authority to approve the project. Before long, the project's backers pulled out.

"Five years ago, that idea would have gotten a different welcome," says Forest Service district ranger Sherry Reed, who also lives in North Routt. "But there's too much traffic, too many new subdivisions. Times are a-changing." Rancher Rossi agrees. After selling the easement on his property, he lobbied hard to help establish a permanent funding source so that other ranchers could share his strategy. "The word 'easement' is sort of controversial," Rossi concedes, "but if we get enough of these easements, we'll be able to keep our kids on the land. I'd hate to think I'll be the last rancher in Routt County."

Thanks to the innovations of people like Rossi and Fetcher, a healthy and effective coalition now exists in the valley among ranchers, conservationists, and local government. Between the early Catamount proposal and the chopping-up of rural lands for second-home development, conservation and agriculture have found a common enemy. The agricultural community has felt particularly under siege, facing a declining market and a squeeze from development. But the valley's heritage runs deep. Ranchers first settled the area's rich meadows a hundred years ago; today, the county ranks number two statewide in grass hay production, and agriculture generates average annual revenues of $20 million. Compared to tourism's $125 million contribution, that may seem insignificant, but the area's resort economy capitalizes heavily on the local ranching heritage.

"Our ranching heritage is as important to this company as snow," says Gary Mielke, president of Steamboat Ski and Resort Corp. "Our unique landscape ranks very high on visitors' lists, so the question for us is, how do we keep it?"

Not just marketers but the community as a whole feels emotionally, or at least sentimentally, attached to the valley's hay bales, its tractors driving through Main Street, and the cattle drives that still occasionally clog the highway. "Saving ranching is very trendy right now, and it's bringing people together," says Mucklow. Indeed, the state's Nature Conservancy recently bought the historic Carpenter Ranch near Hayden, 25 miles west of Steamboat, which it is keeping operational while also helping protect the Yampa River system. In an effort to maintain a credible relationship with valley ranchers, the Conservancy's Jamie Williams set up a local advisory committee to oversee its work in the valley. "Chairing that committee is the most exciting thing I do," says rancher Fetcher. "The Nature Conservancy does an outstanding job of making us feel valuable and responding to our concerns. If they extend an olive branch to us, we should extend one to them."

The Carpenter Ranch symbolizes a meeting of two worlds: the traditional ranching community and a modern science-based approach to conservation. The Nature Conservancy operates the 1,000-acre cattle ranch with an eye to the innovative spirit of Ferrington Carpenter, the late esteemed Hereford bull breeder and first director of the U.S. Bureau of Grazing in 1934. Carpenter died in 1980. For thirty-nine years, Mel Williams managed the ranch for the Carpenter family. "Ferry was a wonderful man," says the fourth-generation rancher. "He was progressive in his operation and brilliant at marketing his bulls, which were famous throughout the West." Williams agreed to stay on the ranch and help the Conservancy through its first year of operation.

"We intend to carry on the tradition of innovation at the ranch," says the Conservancy's ranch manager, Geoff Blakeslee. "It will provide an opportunity for ranchers and conservationists to learn about each other and achieve a common goal of a sustainable working landscape in the Yampa Valley."

Such goodwill has traveled far, helping pave the way for a government-backed, citizen-initiated Open Lands Plan. Its broad mission was to come up with widely accepted tools for conserving ranch lands and natural areas.

Of the plan's strategies, the most promising so far has been the Purchase of Development Rights program, know as PDR. The citizen-driven PDR initiative to increase property taxes passed by only 94 votes in a county of 7,000 voters, underscoring the ambivalence many westerners still hold toward land-use programs. What made the difference, however, were the PDR program's strengths: It is a voluntary, market-driven approach that keeps land in private ownership and on local tax rolls. It also acknowledges the costs of keeping land open for the public good, a point conservatives have been driving home for a long time.

The money, about $360,000 a year from a small tax increase, will help leverage additional state and federal dollars to provide ranchers with one more voluntary tool to help them stay on the land. With money they earn from selling permanent easements, they can reduce their debts and avoid selling their property. As one of only a handful of similar tax-approved PDR programs in the country, it will serve as a model to numerous other rural communities grappling with similar issues. Before the initiative passed in

Steamboat, local officials gave the concept a boost by helping fund a demonstration project to purchase easements. The city and county each contributed $125,000, matching a $250,000 grant from state lottery funds administered by Great Outdoors Colorado (GOCO). For the Rossi deal, the county supplied an additional $55,000, and GOCO an additional $300,000.

The county's progressive open-space policies have proved positively magnetic to outside funders. Even the national 1996 Farm Bill contributed to the effort, awarding $430,000 to yet another county-backed proposal to purchase easements by the Yampa Valley Land Trust and the local Nature Conservancy. "Of all the things we do, I'm most excited about the things we've been able to accomplish around conserving our open lands," says Routt County Commissioner Nancy Stahoviak.

By far the biggest feather in the region's hat, and the biggest money earner, is the Yampa River System Legacy Project. In 1996, GOCO, with its hefty lottery coffers of $18 million per year, awarded six ambitious "Legacy" grants that "address regional needs, include multiple public and private partners, and integrate at least two of four funding areas of outdoor recreation, wildlife, open space and local government." After years of partnership building and bold open-space initiatives, the Yampa Valley stood ready to reap its reward.

As part of its application, the valley pulled together a diverse committee that included government representatives from the towns of Steamboat, Hayden, Yampa, Oak Creek, and Craig, as well as Routt County and neighboring Moffat County, the U.S. Forest Service, Bureau of Reclamation, Colorado Division of Wildlife, Colorado Division of Parks and Outdoor Recreation, private landowners and business owners, conservation groups, the Steamboat Ski and Resort Corporation (now owned by the American Skiing Co.), and the Steamboat Springs school district.

Under the project's general theme ("To protect and enhance the ecological health of the Yampa River and the productive agricultural lands it supports while providing for appropriate recreational opportunities"), the coalition defined five major goals: Protecting 5,000 to 7,500 acres of riverside habitat and ranchland through voluntary conservation easements and other agreements; managing recreational use at existing river access sites; integrating public access and conservation on several miles of intensely used river frontage; helping recover native cutthroat trout in order to prevent its listing as an endangered species; and educating the public about the river valley's unique natural and cultural heritage.

When GOCO awarded the Yampa River System Legacy Project $6 million in 1996, it was its second-largest grant ever in the state, and by far its largest per capita. "The kind of partnership that developed in the Yampa Valley is rare and is very much a model for the rest of the state, as well as the country," opined Will Shafroth, executive director of GOCO. "The lesson is to always broaden your stakeholders, and be committed."

Even local government officials once unmoved by conservation grew elated. "I always appreciated the aesthetics of the valley," says city council president Bennett, a jolly furniture salesman with a handlebar mustache. "But it wasn't until I realized how important and unique it was through the eyes of groups like The Nature Conservancy that I got ambitious. Now I know it's possible to dream the big dream about conserving the place." Others in the community have grown ambitious in similar ways. Even the developers of Catamount—once the most controversial land use in the area—have now turned to conservation options for the property.

"It's really remarkable what's happening in this valley," continues Bennett. As he speaks, he leans back comfortably against a public bench. He looks out into the lively riffles of the Yampa River, and then up and down the adjacent Yampa River Core Trail, part of the city's $4 million urban greenway and parks program. He looks pleased, pleased with the day's sunny weather, pleased with where the Yampa Valley is heading.

"We have the passion," Bennett continues. "We like being on the cutting edge." And then he leans forward, about to offer me his newest favorite expression. "Good planners," he grins, "make great ancestors."

Wild Olympic Salmon: Art and Activism in the Heart of the Dragon

Fran Stefan

Long ago when every stream was full of sea dreaming salmon, on an island in Anderson Lake [in Washington State's Olympic Peninsula] there lived a dragon named No-Qui-Los by the ancient people. The dragon's name meant "the demon" and the old people kept safe distance, though even from far away, they could see the golden glint of scales and iridescent purple eyes of the dragon sunning on Tomanawos Rock.

—*No-Qui-Klos, the Dragon of Tomanawos Rock*

Mall Johani stared at the map; for the hundredth time she studied features of the Olympic Peninsula on the northwest corner of Washington State. The massive Olympic Mountain range forms a domelike core of this landscape, with spires reaching heights of 8,000 feet and then falling away into valleys of old-growth forests, rain forests, and rivers in every direction. To the west, north, and east, the peninsula is bounded by salt water. And it is the Pacific salmon's homeward migration to the peninsula's inland streams and rivers that defines life for many who live here.

On the inner edge of the peninsula, clean-swept coastlines give way to a bulbous outcropping called East Jefferson County, which now drew Mall's attention. Her artist's eye saw in the landmass the outline of a dragon's upright shape. To the north, the ocean currents from the Strait of Juan de Fuca curved around the dragon's arching neck and craggy head and then flowed in a southerly direction into Puget Sound. Saltwater inlets, bays, and islands etched the dragon's belly, legs, and tail, while its vast wings arched backward into the deep valleys, forests, and rivers of the Olympic Mountain foothills.

And Mall's home, the town of Chimacum, lay at the heart of the dragon.

Journey

> One day a young Chimakum warrior named Quarlo decided to seek out the sun bright dragon and learn its secrets and perhaps capture its power.

Mall's fascination with the maps of the Olympic Peninsula was not idle curiosity. Rather, she sought an image to guide Wild Olympic Salmon, a new community effort that Mall, her husband Tom Jay, and a few close friends founded in 1987. Mall, a brilliant artist in her own right, seemed to have touched sketch lines from a deeper past. For, to her amazement, barely two weeks after she discovered the dragon in Jefferson County, the myth of *No-Qui-Klos, the Dragon of Tomanawos Rock,* fell into her hands.

No-Qui-Klos is the sole surviving tale from the Chimacum Indians, as told by La-Kah-Nim, a S'Klallam Indian Chief of the Olympic Peninsula known for his regal bearing and generosity toward settlers. The story takes place in Anderson Lake at Tomanawos Rock, just west of Chimacum, and it speaks of a time when powerful natural forces and mankind held one another in friendship.

Ultimately, *No-Qui-Klos, the Dragon of Tomanawos Rock,* became the cultural soil for Wild Olympic Salmon, grounding that group's efforts to protect wild salmon in a sense of place, history, and Native American lore. Equally important, the magical No-Qui-Klos is a potent environmental symbol. *No-Qui-Klos,* as rewritten by Tom Jay, evokes a sense of reverence for the power, mystery, and wonder of the natural world in which salmon live. Thus, the vaulting image of No-Qui-Klos runs like a silver thread through many of Wild Olympic Salmon's projects—weaving together the myth, the salmon, the hydrologic cycle, and the watersheds of East Jefferson County.

Friendship

> No-Qui-Klos hissed, and glided towards Quarlo, hypnotizing the warrior with its dragon gaze. Quarlo closed his eyes as the dragon's tongue flickered about him, but to his aston-

Olympic Peninsula

ishment, the dragon did not devour him, but began to caress
him with dream-like dragon kisses. Quarlo stroked the
dragon's golden head, and No-Qui-Klos lay down by his side
as if they had been friends for life.

Led by the artistic team of Mall and Tom, the early days of Wild Olympic
Salmon, 1987 to 1992, were a panorama of educational activities that celebrated the salmon resource. Under their leadership, Wild Olympic Salmon's
projects were expansive, technically complex, and infused with a rich sense of
vitality, color, texture—and wry humor. The hallmark of their work has been
the ingenious use of art to raise sobering environmental issues to the octave
of hope and enchantment.

Take, for example, Fin (*Oncorhynchus bodacicus giganticus*), Wild Olympic
Salmon's trademark fish—25 feet of fiberglass, painted a silvery sheen to
match the look of Pacific salmon, and big enough for children (and a few willing adults) to crawl into her belly. Fin, so-named by local schoolchildren,

travels across the United States carrying the message that migrating salmon and their watersheds are worth saving. Fin's belly is interesting; she carries a watershed in it. Her interior is painted to depict a natural forest setting with a stream running through it, the salmon's freshwater habitat. Fin has more than 150 plants and animals that share her watershed with her—painted on her insides. The beguiling Fin has turned heads at universities, state fairs, local parks—even gas stations—and she has appeared in more than fifty newspaper, television, and radio stories. Her method is playful, engaging; her message is deadly serious.

Another Wild Olympic Salmon keynote is its biennial Salmon Festival, which coincides with the return of migrating salmon. Mall orchestrated the first festival in 1989, and, as she put it, "the issue was the wonderfulness factor." For days prior to the event, locals rushed to build the booths, decorate the Community Center, raise the big Centrum tent, dig the salmon baking pit, rope off the parking lots, and attend to a thousand details that would transform Chimacum Park into the Salmon Festival on Thanksgiving weekend. The festival was vibrant with colorful costumes, singing, dancing, storytelling, and a salmon feast. The first year, Wild Olympic Salmon presented members of the Quileute Tribe with a huge hand-carved dragon bowl as a token of the group's respect for those who had loved and respected the Chimacum watershed for thousands of years.

Deborah Michel, current director of Wild Olympic Salmon, describes her experience of the Salmon Festival this way: "Our biennial Salmon Festival is the epitome of community celebration to me. . . . It is full of music, art, and poetry inspired by salmon. It gets dark early in late November when we hold the festival—that's when the salmon return. After dark it becomes magical, the long fire for barbecuing the salmon, campfires and bonfires, luminaries lining the path through the woods to the big tent."

The enchantment of the Salmon Festivals has included colorful, exotic games that intrigue children and adults. One of the first was the Watershed Trading Cards—the naturalists' version of baseball cards. With a grant, Wild Olympic Salmon hired twenty artists to design forty cards, each a unique depiction of the flora and fauna of Jefferson County's watersheds. The cards are a feast for the eyes: red cedar, western hemlock, Douglas fir, spruce, willow, ferns, eelgrass, red-tailed hawk, falcon, osprey, eagle, heron, kingfisher, black bear, deer, elk, seagulls, coyote, raven, and salmon.

The backs of the cards are actually puzzle pieces forming the "Heart of the Dragon" map of East Jefferson County that Mall had labored to draft. The trading cards first became available as a contest for children and adults. Some 10,000 cards were printed and sold locally in rainbow-hued envelopes of 4 cards for a dollar. The object of the contest was to collect all 40 cards, arrange them to form the complete watershed map, and study the map for clues to a few "mystery questions." Contest winners received handcrafted silver salmon whistles.

Recalling the first Salmon Festival, Mall chuckles: "You could see kids and families, sitting outside on blankets, swapping cards with one another— 'Don't you have the frog?' Some people got fanatic about it! The trading cards introduced the whole thing with the dragon and the watershed."

By far, one of Mall's most ambitious and original undertakings was a magical game, Tracking the Dragon. In 1990 she outlined the shape of a large dragon track, 2 feet long and half again as wide. A team of thirty-five writers, sculptors, and artists designed twelve of these tracks, cast them in bronze, and placed them on public lands throughout East Jefferson County. Each track depicted an important step in the hydrologic cycle—sedimentary rock, wetlands, rain, clouds, rivers, streams, estuary, glaciers, soil water, groundwater, Puget Sound and Hood Canal, and the sea—the movement of water through the earth and its atmosphere.

To participate in the game, it was first necessary to find a series of clues. Initially, the clue pages were published once a month for a year. Individuals, families, and groups then set out to find the tracks. Some were relatively easy to find; others were much more challenging. Clues were found in books, periodicals, and maps in the library. To win, players had to find the footprints and describe the images sculpted on them. Thus, each winning team or player took at least twelve hikes through some of the most beautiful and interesting areas in the Olympic Peninsula.

For example, the first footprint can be found "down the thread of the sea" on the granular surface of sedimentary rock on the high tide mark of Indian Island, to the east of Chimacum. To find it, dragon trackers cross the bridge to this small island and then follow the downhill path to the Lions Park shoreline. Here, they creep over boulders overhung with brush, cross sandbars white with shells, and follow the margin of the sea to discover the bronze dragon track embedded in sedimentary rock. At dusk, the track comes alive as the slanting rays of sun strike the intricate designs of the bronze footprint.

Dragon tracker Dennis Kelly of Port Townsend, Washington, captured the essence of the game when he said: "What I have enjoyed most about Tracking the Dragon is the opportunity to become, not childish, but childlike, once again. For a brief period of time, the distinction between adults and children has blurred, and we're all, the eight of us, childlike friends pursuing a year-long gigantic scavenger hunt. We've been able to rekindle our sense of wonder, spend some good times with our families and friends exploring this wonderland in which were live, and we've all created for ourselves a tapestry of rich experiences in the process."

Wild Olympic Salmon compiled the dragon clues, many of which were original maps, stories, poems, and songs of this area, into a beautiful book entitled *Tracking the Dragon*. Some of the book's most lyrical passages are the journal entries for Snow Creek, taken from a local family diary, *Stream Book*. Beginning with the family's first visit to the creek in 1889, they end with this entry in 1944: "Fifty-five years by the stream, studying and learning some of its secrets . . . the last of its secrets revealed to me . . . through the discovery of a fossil track."

Adventures

> Quarlo and No-Qui-Klos had many adventures. They fol-
> lowed the salmon to their villages under the sea where
> salmon lived as people like you and me. With No-Qui-Klos
> as guide Quarlo learned to understand the wind's whisper
> and raven's raucous riddles.

In 1992 Wild Olympic Salmon shifted focus as members rallied to save local stocks of wild salmon from extinction and repair damaged watersheds. Here the group applied its ingenuity and command of scientific principles to the hands-on world of salmon enhancement.

Chimacum Creek drains the watershed that is Port Townsend, Port Hadlock, and Chimacum Valley. In the late 1980s a logging road washout poured tons of debris into the creek channel and destroyed the remains of the indigenous wild chum salmon run, a fish almost sacrosanct to many who live in the heart of this valley.

At first, members of Wild Olympic Salmon did not understand how seriously Chimacum Creek was damaged. In the fall of 1989 they set up a weir in Chimacum Creek to count the chum salmon returning to spawn in the lower reaches of the estuary. Debra Michel, a Wild Olympic Salmon volunteer at the time, remembers it this way: "We set up a weir to count the indigenous wild chum of Chimacum creek. It was lovely, we worked on weekends, in small groups, by the creek, in the creek. It was a wonderful time of camaraderie. When the weir was finished we took turns watching it. For days we rotated watch, we wrote poetry on the wooden weir itself, a journal of data and poetry was kept. It had become a sacred space.

"The first year we counted no chum. We were undaunted, we must have done something wrong, we said, they must have gotten through the beavers tunnels, we must have missed them, they just have eluded us. We took the weir out and put it in the creek again the next year. This was no small task, carrying heavy wooden weir pieces up a steep trail, a long way to the road. By the third year we had only seen one chum. Essentially this meant they were extinct after living here for 12,000 years. Things did not seem so lovely now. We mourned their loss for a long time, not knowing what to do."

Four years ago, Tom Jay and a cadre of volunteers forged a team effort with the Washington State Department of Fish and Wildlife to bring the wild chum salmon back to Chimacum Creek. Ultimately, they decided on a plan to stock the creek with eggs from the neighboring Salmon Creek. This method, if successful, would maintain the gene pool essential to wild salmon populations. But Salmon Creek had troubles too—the number of returning salmon had diminished to a critical level.

Fall days, the lower reaches of Salmon Creek are alive with splash and movement of small stones. It is the sound of the chum salmon, returned home after three years at sea to spawn and die. During migration, the salmon have fasted and used muscle protein in order to gain sufficient energy to complete their travels and the ensuing spawning process. Now, they swim in rapid zigzag motions across the shallows before finding the spot where females will dig holes in the gravel for their nests, while sparring males keep a watchful eye. And all around are the carcasses of spent salmon, whose nutrients will eventually find their way deep into the surrounding soils or even to the tops of trees that line the stream banks.

Chum salmon (*Ochorhynchus keta*) ascend rivers only a short distance to spawn, then the young drift to the ocean immediately after emerging from the gravel nests, or redds. Normally, only about 30 to 40 percent of the chum eggs will ever hatch, and when the young do emerge they are vulnerable to predation by other fish. When stocks have been severely stressed, as they were on the Salmon Creek in 1990, survival rates of the young fry are even more precarious.

To increase the survival rates, Wild Olympic Salmon designed and built a hatchery on Salmon Creek. It's a one-room building with wild colorful salmon emblazoned across the stoop. The small interior is filled with hoses, pipes, tubs, and trays for nursing salmon eggs into fry. Tom Jay explains: "Our process is to take 20 percent of the eggs from females that come in to Salmon Creek before, during, and after flood events—this is the way fish naturally hedge their genetic bet. We duplicate that bell curve, replicating the genetic diversity, and the egg-to-fry survival rate is high, and it builds the population a little faster than we would otherwise."

Once the salmon eggs are fertilized, the Washington State Department of Fish and Wildlife holds the eggs in its Dungeness hatchery until they have reached the eyed stage of embryonic development. Later, the eggs are moved to the Salmon Creek hatchery, where they are tenderly hand-reared by Jay and his crew until spring, when the little fish are moved to saltwater net pens in Discovery Bay. Once the fry have grown to about 3 inches long, they are released to the open sea with a grand Wild Olympic Salmon celebration.

"When the chum fry are released into Discovery Bay in the spring," Deborah Michel writes, "they are sent off and wished well by a large group of people. There is a great sense of community, satisfaction, and hope. We sip champagne and toast our chums, we read poetry we have written for the occasion, and we stay out on the dock long after the last fry has been seen."

By fall of 1996, the careful, dedicated expertise of the hatchery crew finally paid off. The stock of chum on Salmon Creek had increased fourfold. Now, with the Salmon Creek population healthy, 70,000 eggs were moved to Chimacum Creek. If all goes well, chum salmon will again migrate homeward to the lower reaches of Chimacum Creek early in the new century. Some speculate the National Marine Fisheries Service's recent decision not to list these fish as endangered is largely because of the success of this project.

Tom Jay is a well-known Pacific Northwest poet and sculptor. When asked how he balances raising salmon and art, Jay answers evenly: "The way it is for me is that they are all intertwined. I just stay in the salmon; hatchery work is the most natural in the world. Art-in-life, life-in-art."

In 1994, Wild Olympic Salmon again enlarged its scope to include hiring seven displaced timber workers under Washington State's Jobs for the Environment Program. The largely state-funded program is a response to the cutbacks in old-growth timber harvests in the Pacific Northwest that had resulted in 20,000 unemployed timber workers by 1992. The men and women in the Jobs for the Environment Program work a forty-hour week, six months a year, at wages comparable to those of their previous timber industry jobs. But Jobs for the Environment gives timber workers a new profession: restoring damaged watersheds.

Jobs for the Environment is a true partnership effort. Initially funded by the Washington State legislature in 1993, the program is now managed by the Washington State Department of Natural Resources (DNR). Interested groups—tribes, counties, cities, state agencies, and nonprofit organizations—can apply for a grant. Since the program's inception in 1993, DNR has funded 90 projects covering some 250 sites. Typically, once a grant has been awarded, a committee is formed to determine site priorities, timing for projects, financing, and technical needs.

William Michel leads the Jobs for the Environment program for Wild Olympic Salmon. The seven displaced timber workers, ranging in age from thirty to fifty years, that Michel hired in 1994 are known as the Quilcene-Snow Restoration Team. Crew members work on several watersheds bounded by Snow Creek to the west and the Quilcene River system to the south, or about a 318-square-mile area within East Jefferson County.

Says Michel, "We provided crew members training in the life cycle of salmon, hydrology, geology, plant identification classes, fish habitat, the basics of ecology, environmental principles, and the impacts of logging and agriculture on watersheds. But when we brought out big machinery [for some of the larger restoration projects], the Jobs crew was like kids in a candy store. Their eyes sparkled—they *knew* how to use this machinery—prior to that they had a steep learning curve. The most important challenge for me was making sure they knew how to do the job right; if they knew how to do the job right, they did.

"Most of these guys came from timbering communities; their neighbors were loggers. As the year progressed, a schism developed. Crew members were trying to heal wounds from faulty timber practices, and yet they still lived in their home communities. They began to see certain practices need to change. For example, trees are the backbone and skeleton of the stream. Remove that and there is nothing to hold the stream in place. The crew began to feel all loggers should have training in the importance of protecting riparian zones."

By 1996, the pristine Snow Creek chronicled in *Stream Book* suffered from a broken back. Trees had been logged on a large tract of public lands adjacent to the estuary. Without the skeleton of the trees' root systems to hold the soil in place, landslides occurred, pouring a tremendous load of sediment into the Snow. The sedimentation eventually raised the creekbed by several feet, causing repeated flooding downstream by forcing water out of the debris-filled channel. The Jobs crew excavated sediment and anchored larger woody debris to lower the streambed. Essentially, they created an artificial skeleton in the stream—to hold it in place and make the pools and riffles that are a part of a healthy environment for summer chum salmon.

Michel sees other possibilities for the Quilcene-Snow Restoration Team. "At the end of this grant on June 30, 1997," he says, "the crew members will have worked two and a half years and will have become quite skilled. This stream work is not an art; it is a science. Much of it is intuitive and requires a lot of problem-solving ability. We hope to transition the crew into Jefferson County Public Works."

Passages

> But Quarlo was mortal, and not even No-Qui-Klos could save him from death. When his great grandchildren were grandmothers and grandfathers, Quarlo died and his spirit skipped across the jumping log to the land of shadows.

By 1994, leadership within the Wild Olympic Salmon organization began to shift. As in so many environmental programs spanning several years, the baton must eventually pass to a new team of individuals to carry it forward. For everyone involved, the transition can be difficult, even painful.

Deborah Michel, initially drawn to this organization because of its life-giving spirit, now holds the position of Wild Olympic Salmon Director. At times, she finds it a dance. "Picture standing in a circle," Deborah says, "you back up because you have just stepped on someone's toes and as you back up, you step on someone else's toes." The new members on the Wild Olympic Salmon Board of Directors find themselves equally challenged to cast their own imprint and direction on the organization.

To be certain, Mall Johani and Tom Jay cast a larger-than-life shadow. Beyond their visionary leadership of Wild Olympic Salmon, their artistry has given Chimacum valley an identity found in the symbology of the dragon and the fish.

Today, Chimacum Park is graced by the magical presence of Mall's life-size sculpture of No-Qui-Klos, the Dragon of Tomanawos Rock—even as his large footprints lie embedded in the watersheds of this area and his story rises out of the lore of the land.

Similarly, Tom's sculpture of "Heroic Chum" sits in the business center for the communities of the Chimacum valley. The bronze sculpture is the head of a summer chum salmon, 7 feet tall from gill cover to snout, about 3 feet wide, surrounded by a pool of quiet water. According to Tom, "Originally I wanted to break up the pavement, have the salmon erupting through it—but the contractor wanted something more sedate. It would have been more fun to have it wild . . ."

> When Quarlo died No-Qui-Klos was disconsolate. Try as they might the villagers could not comfort the sorrowful dragon and when they laid Quarlo in the death canoe No-Qui-Klos plunged in the waters of "Wuldge" (Puget Sound) and disappeared, though some claim to have seen him off Kala Point in the early morning when Dog salmon run.
>
> So that is the story of the No-Qui-Klos, the Dragon of Tomanawos Rock, and how its love of a young warrior changed the face of where we live . . .

Oregon's Plan for Salmon and Watersheds: The Basics of Building a Recovery Plan

Carlotta Collette

Thirty years ago, I lived for a short time in the Coquille Valley on Oregon's south coast. It is a wet place. The one winter I lived there, water was everywhere. While I watched, the broad, flat bottomland disappeared under a new lake. For farmers and ranchers trying to work that floodplain, this much water was something to contend with. People dug canals to drain it off, put in tide boxes to keep it out, and dumped whatever was handy onto the valley floor to raise the earth up. Most of this behavior had long been promoted by government policies touting reclamation of the land for "production."

But for the fish that share this valley's waters—especially Oregon's coastal coho salmon, one of the state's first fish to be listed under the federal Endangered Species Act—the annual inundation meant survival. Left alone, the lower reaches of the Coquille Valley are filled nearly every winter by the tides and by the heavy rains that let loose there on the sodden side of the Coastal Mountain Range. To coho salmon, which spawn far upstream, the maze of seasonal sloughs and slackwater marshes provides the perfect habitat for a winter's rest stop before heading out to sea. The extra time to feed in fresh water makes the fish stronger and more likely to survive their long ocean migrations.

Unfortunately for the fish, and the fishers who relied on them for their economic well-being, few people understood even thirty years ago the relationship between the Coquille Valley's annual flooding and its abundant runs of coho salmon. It was only when things became noticeably wrong—when water quality and the life it sustained were threatened, when salmon populations crashed and fishing had to be curtailed; and when the federal government came back to the watershed with the threat of a different set of policies—that the people of the Coquille Valley got together to care for their river and the watershed it drains. When they did, they got the attention of the state's governor, John Kitzhaber, and set an example that could become a model for how federal and state bureaucracies can work effectively under the guidance of local communities.

Working the Watershed

Paul Heikkila, an agent with the county extension office who's been working with local landowners in the Coquille watershed for more than twenty years, puts it this way: "We're on the edge of the sea here. In the intermountain regions, there's an attitude about things being better over the next ridge. For us, there is no other ridge. We're 'ground zero' for endangered species," he says. "Whether it's the spotted owl, the marbled murrelet, the salmon or you name it, we've had to deal with it here. Our community has suffered both from the loss of our livelihoods—in forestry, fishing, ranching—and from the fear of more federal control here."

Governor Kitzhaber knew firsthand about the suffering in the Coquille Valley and other resource-dependent counties in his state. The timber industry has long been one of the mainstays of his hometown economy. The Coquille Basin's highlands are all in commercial forests, the same forests that surround his old stomping grounds. The governor's friends and neighbors include environmentalists and loggers, fishers, and farmers. He's an avid angler himself. So, when it appeared likely the federal government would declare the coastal coho endangered, the governor was willing to bet that together the state of Oregon and its citizens could craft a better recovery effort than any federal agency could. The model he turned to was in the Coquille Valley.

In January 1995, soon after Kitzhaber was elected governor and about the time the National Marine Fisheries Service was proposing to put the coastal coho on the federal Endangered Species Act list, he accepted an invitation to tour the 1,059-square-mile Coquille Watershed to see what local river basin stewards—the townsfolk, ranchers, timber companies, and farmers—had been able to accomplish there.

Roy Hemmingway, the governor's point man on salmon and energy issues, remembers Kitzhaber returning from his tour filled with enthusiasm. "They've got about a thousand people signed up in the Coquille Watershed," the governor told him. "Something's really going on down there." Kitzhaber

next sent Hemmingway down for a tour himself. "I had my eyes opened," Hemmingway admits.

"People Relate to Their Watersheds"

"Some of us had been working with various state agencies on stream enhancements for more than a decade," explains well-named Charlie Waterman, president of the Coquille Watershed Association. "But we really got going in a coordinated way in about 1993, when a lot of watershed groups were forming around the state. Some people didn't want to come to the table. They were pretty upset at the federal government and suspicious. They figured the government always had more power at the table than they did. But once we got there, they could see that the landowners had as much power as the state and federal agencies did. It turned out to be a lot of fun, really. We got to see what we could do together," he remembers.

Charlie and his family have been watching the Coquille River swell out over the land and shrink back into itself for generations. Between them, Charlie and his father have about 2,500 acres of ranch and forestlands. They raise cattle and sheep, and maintain a small timber holding of about 450 acres.

"We're not 'cut and run' people," Charlie explains. "We've always been here. We know we can't destroy our water or our soil. There's a limited amount. This is pretty basic stuff."

Having "always been here" made it easier for Charlie and Paul Heikkila and the rest of the watershed association members to identify which streamside recovery projects would make the greatest impact first—whether for scientific or strategic reasons.

"The first thing we did was daylight the creek running right through the town of Coquille," explains Heikkila. The "Fourth Street Creek," a tributary of the Coquille River, had long been diverted through a 1,000-foot culvert that proved nearly impassable to salmon swimming toward spawning grounds above it. "Opening the creek was a great attention-getter in town, and young salmon were spotted upstream in a matter of months," Heikkila adds.

Identifying which people needed to be involved was equally important. "The trick is finding the leadership in a community," Heikkila notes. "If you don't have those critical first people, those first visible projects, you won't get very far."

"Heikkila is a genius at this," praises Hemmingway. "It may look like what we sometimes call 'random acts of kindness,' and it may be that somewhat. But you've got to let that local process work. You've got to let those local folks make the determinations about what are the keys to unlocking community support. Some projects may be less important in terms of salmon recovery or watershed health. But they are more important because they are right on the road where everyone will see them, or they're being done by just the right landowner to gain everyone else's trust.

"No way could we in the governor's office in Salem have directed this process. We couldn't possibly have figured out what would work in each watershed across the state. We'd be years behind."

Hemmingway says he doesn't remember exactly who had the idea for knitting together the state's grassroots watershed councils with the efforts of state agencies working on fish and wildlife to try to craft a cohesive restoration effort for the salmon.

"It was clear Oregon had a lot going for it that other states lack," he says. "We had our new forest practices and riparian habitat rules that were just beginning to take effect; our land-use planning that no other state has; our process to respond to the Clean Water Act; our minimum river flows protection, and a lot more. All of these pieces could be brought together with the watershed efforts around the state to constitute a recovery plan."

But the federal government had never accepted a state-proffered recovery plan as an alternative to listing a species as threatened or endangered. For that matter, the federal government has yet to adopt any final recovery plan for any listed salmon population. Hemmingway and the governor thought Oregon's could be the first.

"Our goal was to save the salmon," Hemmingway says. "I'm not sure we thought a 'no listing' was an option. But once we had people working together, some of them—some key stakeholders—felt that avoiding a listing was important. We thought we needed it to gain their confidence."

Making a Puzzle into a Plan

The governor started by calling together his staff, leading fisheries scientists, timber industry representatives, environmental groups, members of local watershed associations, and others with an interest in the salmon. The basic elements were all spread out like pieces to a puzzle, a very big, very complex puzzle.

Some of the pieces were shaped like river basins, the Coquille being one of the models. A growing number of watershed-based community groups were developing local recovery plans and working along their own streams. These groups would recommend stream-reach by stream-reach measures necessary to correct more than a century of land and water management practices that had contributed to the salmon decline.

Other pieces looked like state agencies: the Department of Fish and Wildlife, Department of Transportation, Land Conservation and Development Commission. Whatever they had done on their own in the past had failed to sufficiently slow, stop, or reverse the decline of the salmon populations, let alone rebuild them. In many instances, policies and practices of one agency conflicted or undermined those of another.

"I remember almost a fistfight between two folks from different state agencies over how to proceed," recalls Charlie Waterman about the early Coquille watershed meetings. "We had to get them to agree." The governor took the same approach.

"Every other week the governor met with his agency heads to get them working together," Hemmingway recalls. "Sometimes he had to knock heads. He was very much in the thick of it." The promise was that maybe working together, with the local watershed groups, they could turn things around this time.

Timber and other industry representatives, environmental groups, and scientists also were puzzle pieces with special concerns. Timber companies came forward admitting more culpability for their role in the salmon decline than they would have in the past. To avoid a federal listing, they were ready to try to make amends. Fisheries scientists, bolstered by the ultimatum posed by federal control, saw in the governor's process the opportunity to have a stronger voice than ever before.

At times, the environmental groups looked as if they might leave the process, becoming missing pieces without which the puzzle could never be completed. Some supported the governor's process and stayed with it through the long drafting and review process. Others were wary from the start. They were not convinced that a voluntary recovery plan would carry the same weight as one mandated by the federal government.

Federal agencies had their own puzzle pieces scattered among the others. Old Soil Conservation Districts, Bureau of Land Management offices, Fish and Wildlife Service staff had to face the same reality as the state agencies. Whatever they were doing, it wasn't working. Whole runs of salmon were disappearing, going extinct on their watch. Most of the federal agencies were willing to risk trying a new approach.

To assemble from all of this a recovery plan that stood a chance of rebuilding salmon runs, let alone gaining federal acceptance, the governor and his staff, and everyone else he could bring to the cause, would have to see the puzzle pieces as a whole picture and work diligently to fit them into place.

And they did it.

As the plan began to take shape, several of the environmental groups did pull out, but the governor kept enough of them engaged to help pull the picture together. Then he took the draft out to communities and experts across the state, soliciting their opinions. The folks in Coquille remember "marking up several versions," before they saw a final one that earned their support.

Gaining approval and funding from the state's legislature was another struggle. Oregon's reputation for environmental pioneering was built during years when there was a Democratic majority in the legislature. The legislature in the past decade has shifted more to the right, and the Republicans in charge have demonstrated their intent to dismantle many of the environmental efforts that gave the state its "green" reputation. Nonetheless, with the solid support of the timber companies, the governor was able to win over the legislature, as well.

Finally, even the National Marine Fisheries Service, which oversees the Endangered Species Act where sea-going fish or wildlife is concerned, accepted the plan. In April 1997, the federal agency agreed to forgo listing the coastal coho. The Oregon Plan would have its shot at restoring the runs.

Oregon's was the first state recovery plan the federal government had ever approved as an alternative to listing, but the governor's satisfaction was short-lived.

Winning and Losing and Winning Again

The governor and his team were still celebrating their unprecedented achievement when the half-expected lawsuit was filed in May 1997. Some of the state's most credible environmental groups, including the Oregon Natural Resources Council, the Pacific Coast Federation of Fishermen's Associations, and the Native Fish Society, which had abandoned the planning process, sued the Fisheries Service for not listing the coho. They said the agency had abrogated its responsibility by assuming the Oregon Plan would be sufficient to save the fast disappearing salmon population. After deliberating for half a year, U.S. Magistrate Janice Stewart ruled in their favor on June 1, 1998.

The crux of the debate had been the plan's reliance on the willingness of timber companies and other landowners to voluntarily do what they had been unwilling to do before. From the governor's perspective, this was the plan's chief strength: the readiness of individuals and companies to make significant changes in how they do business.

The federal government controls only about 35 percent of the forests and grazing land in coastal watersheds, the initial focus area of the Oregon Plan. Private holdings make up the rest. Kitzhaber argued, "The standard under the Endangered Species Act for private landowners is basically that you can't do anything that 'takes' any more of the endangered fish. It doesn't require private landowners to do any real watershed restoration. But the habitat needs restoration, and 65 percent of it is privately owned. The whole plan is based on getting people to the table, giving them ownership, and then helping them go further than they have to go."

The governor figured the Oregon Plan would do more for the coastal coho than any federal recovery plan ever could. "If you look at the Oregon Plan," he reasoned, "it takes about 50 to 60 percent of the woods out of production in the Coast Range. The feds could never get that kind of a cut back. The way you make change is by involving the landowners who are affected."

The environmentalists didn't buy it.

"I like Governor Kitzhaber a lot. But his coho plan does more to protect the timber industry than it does to protect coho habitat," said longtime salmon advocate Bill Bakke, director of the Native Fish Society—one of the groups that sued over the salmon ruling. "We got into this mess because of the voluntary efforts on the part of special interest groups," he added. "It can't be just voluntary."

Judge Stewart had agreed. "Voluntary actions, like those planned in the future, are necessarily speculative," she wrote. "Therefore, voluntary or future conservation efforts by a state should be given no weight in the listing decision."

The governor immediately ordered his state to appeal and asked for a stay until the appeal was ruled on. He was concerned in particular because the state tax on timber activities that he'd counted on to help fund the recovery effort had a "fail-safe" mechanism: If the coho were listed, the tax was terminated the same month. The judge called that "a self-inflicted wound." The stay was denied.

Then the Fisheries Service, under the auspices of the Department of Justice, also filed an appeal and asked for a stay. The judge denied it as well.

On August 3, 1998, Oregon's coastal coho were listed as threatened under the federal Endangered Species Act. Soon afterward, some of the bigger timber companies announced they'd voluntarily continue to support the recovery effort and the governor's plan. They even called for a special session of the state legislature, which had already adjourned, to restore the tax.

Eventually, the state dropped its appeal and moved instead to broaden rather than pull back on the salmon plan. In an executive order in January 1999, Governor Kitzhaber expanded the coastal coho initiative to include protections and restorative actions for all wild salmon runs and watersheds in the state.

The environmental groups that had filed their suit were satisfied that, with the federal listing, the Oregon Plan stood a better likelihood of surviving the governor's term in office to be implemented even in the absence of his leadership. They are not effusive in their support, noting that the plan still needs to go further, particularly regarding forest practices in the state, but their focus has, for the most part, shifted to pushing for full and consistent implementation and monitoring of the effort.

"That's what it's going to take," says Bakke. "It's individuals that make a commitment in their place and get others to join them in that commitment that will make the difference. It's a mistake to think politicians are going to solve the problem. They have to be forced."

Epilogue

It is November as I write this. Three weeks ago, I was in the Coquille Valley watching, as so many years before, the land disappear under a steady rainfall. Paul Heikkila drove me from one creek to another, pointing out the abandoned dairy farm that had been purchased to restore a part of the estuary; the small ranch where the original creek, long buried, was discovered and reopened to let coho back in. "That one was an ideal project," Heikkila brags, "because we needed to repair the road anyway. This saved a lot of money."

All of the work and nearly all the materials came from Coquille or nearby communities. Many of the workers were displaced loggers and fishers. The habitat repair work provides "real family jobs," he explains.

"We're looking at functioning communities, not just watersheds," Heikkila says. "The fish had a real economic value to this valley. Between 1988 and

1998, the value dropped from $12 million to $2 million." For a river basin with a population of about 17,000 people, that's big business and a big loss.

About half the coastal coho in Oregon came out of the Coquille and its neighboring watersheds, the Coos and the Tenmile. There were canneries and freezing operations all along the lower Coquille River. The basin's chinook salmon runs are still healthy, growing by 6 to 7 percent each year. "If the rest of the coast was performing as well as these three streams are, there wouldn't be a coastal coho listing," Heikkila notes.

In the first three years of its existence, the Coquille Watershed Association helped construct 80 miles of fencing to keep livestock out of salmon streams. They replaced or repaired 9 culverts to enable salmon to travel through. They put in the waterways more than 160 structures—downed trees, boulders, and the like—to create resting pools, prevent the washout of spawning gravel, and increase juvenile salmon rearing areas.

Statewide, the numbers are downright hopeful, especially where private land is concerned. The plan's 1999 Annual Report describes the 400 miles of roads on private land that were voluntarily improved in 1998. More than 4,000 culverts were repaired or removed. Fifty miles of logging roads were decommissioned, and 30 miles were temporarily closed to prevent erosion into streams. Hundreds of miles of creeks were fenced off to protect them from cattle and other livestock. Thousands of trees were planted to shade streams and stabilize streambanks. All of this was on private land.

In fact, in the year between the Oregon Plan's adoption by the people of Oregon and its rejection by the federal court, volunteers and local communities in the state completed more than 1,200 watershed restoration projects. Their work has simply accelerated since then.

In the Coquille, as in most watersheds, there are waiting lists of people offering their land for recovery projects. Watershed staff can't put together the funding and the scientific reviews (a recent addition) to approve the projects fast enough. Jennifer Hempel, staff coordinator of the Coquille Watershed Association, describes the change she's seen: "In the early days, most of the landowners I'd talk to would have that look, that 'hands on their hips, I'm not budging' look. They'd say, 'the government requires it, the government can pay for it.' Now more people are enthusiastic. They want to do the work themselves. When they do it themselves, they're more likely to maintain it, to take pride in it, to show it off."

"If these people are successful at the local level in becoming champions of the salmon, they're going to demand that they have fish there in the future," admits Bakke. "State agencies and legislatures won't be able to back down without feeling a big thorn in their sides. The watershed councils will demand the return of their salmon."

But Governor Kitzhaber says salmon aren't really the point of all this.

"This whole thing is about how people in Oregon acquire a deeper level of understanding that how they live their lives affects the environment in which they live. It's about changing the culture."

Charlie Waterman understands. "People relate to their watersheds. They may not relate to their geopolitical boundaries, but they see water flowing by, and they value clean water. This is," as he says, "pretty basic stuff."

Bitterroot Grizzly Bear Reintroduction: Management by Citizen Committee?

Sarah Van de Wetering

In their expedition across the newly acquired lands of the United States in 1803–1805, Meriwether Lewis and William Clark encountered numerous grizzly bears. Lewis described one of the first grizzlies he saw as "verry large and a turrible looking animal, which we found verry hard to kill." Nonetheless, the party killed quite a few grizzlies during their journey, including at least 7 while camped in the Bitterroot Mountains that today form the border between Idaho and Montana. There were as many as 50,000 grizzlies living in what is now the continental United States prior to European settlement.

Today, there are between 800 and 1,000 grizzly bears in the lower forty-eight states, concentrated in a few isolated populations in Montana, Idaho, Washington, and Wyoming. By the end of the nineteenth century, grizzly bears in the Bitterroot Mountains had been decimated by hunting, trapping, and predator control projects. Although persistent reported sightings have led some biologists to conclude that at least a few grizzlies remain in the area, federal scientists are operating on the assumption that the grizzly is completely absent from the Bitterroot ecosystem. The nearest recognized populations are in the Cabinet-Yaak ecosystem in the northwestern corner of Montana and not far to the east of that, in the Glacier National Park–Bob Marshall Wilderness ecosystem.

Grizzly bear survival is complicated by the species' own characteristics. Thomas McNamee, in his book *The Grizzly Bear*, described the bear's survival as "a fragile thing indeed . . . the grizzly's needs intersect (or interfere) with so many of an expanding human population's needs or desires, and like all complex systems it is highly vulnerable to breakdown."

Such a breakdown may occur at any of a number of vulnerable points in a bear's lifetime. According to research summarized in the recently released Bitterroot draft environmental impact statement (EIS), female grizzlies do not mature sexually until they are four or five years old; then they take two years to raise each litter (the typical litter size is two cubs). With such a long reproductive cycle, the loss of a single female bear is a significant event in the species' survival, and population numbers are replenished slowly.

The draft EIS also describes how adult grizzlies range over large areas and typically do not tolerate human development. Any proximity of bear habitat and human development can lead to conflicts, injury, or death—usually to the offending bear, although there is always some risk to humans. Forest roads can be a problem for bears either by discouraging their use of adjacent habitat or by leading to their death from vehicle collisions.

The largely roadless and undeveloped Bitterroot ecosystem offers the opportunity to anchor a new population of grizzlies, potentially boosting the species' total in the lower forty-eight states by as much as a third. Many have pointed out that the Bitterroot ecosystem today lacks the abundant salmon and whitebark pine that once apparently comprised a large part of the Bitterroot grizzlies' diet. But the bear is a notoriously adaptable omnivore, feeding on a wide variety of plants, animals, and insects, depending on seasonal abundance. The Bitterroot draft EIS concludes that, "Most authors . . . agree that successful bear recovery will be determined by the level of human-caused mortality. Grizzly bears can live within the boundaries of the [Bitterroot ecosystem], but their densities will likely be less than what could have been supported when both salmon and whitebark pine were common."

Protection . . . Required by Law

In 1975, the U.S. Fish and Wildlife Service (FWS) listed the grizzly bear as a "threatened" species in the lower forty-eight states after determining that although the grizzly is not yet "endangered" as defined by federal law, it is heading toward that status throughout all or most of its range. Once a species has been so listed, the Endangered Species Act requires the FWS to designate habitat critical to its survival, make sure that federal activities do not degrade the critical habitat, and write and enforce a detailed recovery plan.

The main objectives of the grizzly recovery plan, released in 1982, were (1) identify existing grizzly populations in the lower forty-eight states and establish viable population sizes for each; (2) identify the factors that could limit populations and habitats; (3) propose recovery methods; and (4) achieve

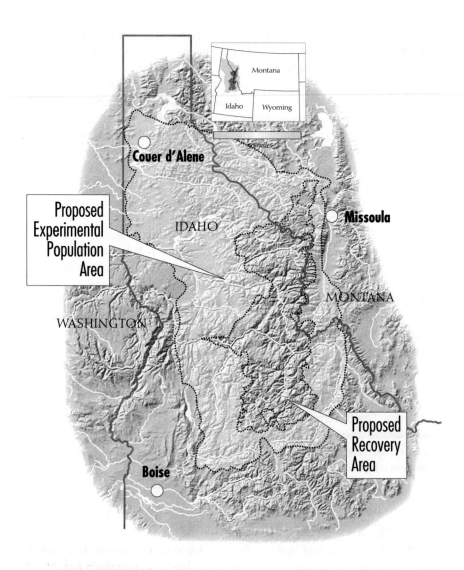

recovery in at least three of the six identified populations. The ultimate goal is to reach a point at which the grizzly can be removed from the ESA list of threatened species. Of the six potential recovery areas identified in the plan, the Bitterroot is the only one not presently occupied by grizzly bears.

Biologists embarked on a number of habitat studies, taking about a decade to start getting specific about how to bring the bears back to the Bitterroot ecosystem. At that point, the FWS prepared a list of preliminary alternatives, including a proposed plan of action for the Bitterroot ecosystem.

According to the plan announced in 1993, the agency expected to obtain twenty to thirty grizzly bears from British Columbia (and perhaps from established populations in Yellowstone or Glacier) and release them into the designated wilderness areas in the Bitterroot ecosystem. After the initial five-year reintroduction period, the grizzly population was projected to grow to a target of about two hundred in forty to sixty years.

Controversy began bubbling up as soon as the FWS began holding public information meetings to describe this proposal and to gather public comments. Dan Johnson, of the timber group ROOTS (Resource Organization on Timber Supply), recalls the public meetings in 1993 as "disastrous." "People were arguing with each other, and each meeting brought new people to start the fight all over again."

Tom France, with the National Wildlife Federation, describes widespread fears that a herd of grizzlies was figuratively poised at the U.S.-Canada border, ready to descend into Idaho and Montana and wreak havoc along the way.

It was not exactly a time of rational discourse.

Getting Beyond "No, Hell No"

After the controversy helped to force the FWS to delay publishing its Bitterroot chapter, some of the key players attended the winter meeting of the Interagency Grizzly Bear Committee at the U.S. Fish and Wildlife Service Regional Office in Denver in December 1993. Dan Johnson was there. So were Tom France, and Defenders of Wildlife representative Hank Fischer.

Dan Johnson remembers feeling like a stranger as he walked through the halls of the FWS office. "Service employees were greeting France and Fischer by their first names. They didn't know who I was. And France and Fischer were pursued by the press. The press showed no interest in my representation." Both experiences reinforced Johnson's perception that the whole recovery effort was being run in cahoots with the conservation groups.

When he had his turn to speak at the meeting, Johnson reiterated the timber industry's position: "We don't want *any* damn bears." But, perhaps feeling the press of opinion in that room, he went on to add that if bears were going to be reintroduced, then groups like his wanted a voice in how it was going to be done.

Tom France and Hank Fischer—veterans of the acrimonious war over wolf reintroduction in the Northern Rockies—heard the "but" in Johnson's comments. They sought him out that evening during the social hour and suggested that there might be some common ground to work from. Warily, Johnson agreed to bring some of his cohorts to meet with them. And thus a rather unlikely partnership began.

One of the early meetings took place in a small conference room at the National Wildlife Federation office in downtown Missoula. "Dan [Johnson] brought all these millworkers, who crowded around our table," Tom France recalls. "It was big labor at its most literal." They didn't make any decisions except to keep meeting. And they did, driving back and forth between Idaho and Montana many times for the next year. They brought others to the table, including the late Seth Diamond (of the Intermountain Forest Industry Association) and Bill Mulligan (owner of the Three Rivers Timber mill in Idaho).

Among the most contentious issues was how to involve local interests in the grizzly management decision process. The conservationists suggested a citizen advisory committee, similar to those already used in other areas of federal land and resource management. The timber folks said no: such committees (in their experience) were often little more than window dressing and seldom influenced policy decisions. They wanted a committee that actually made decisions, which would be carried out by federal agency officials.

This had never been done, so far as anyone knew. That meant there were no models for designing the committee, although state wildlife commissions offer a rough comparison. There were questions about whether it was legal for the secretary of the interior to share management authority with a group of citizens. And the proposal raised all kinds of concerns about who would serve on the committee and under what conditions.

The group pressed on. With the help of attorney Wayne Phillips (counsel to the Montana Department of Fish, Wildlife and Parks), it drafted a detailed plan for Bitterroot grizzly reintroduction. The plan depended on the flexibility made possible by a special provision of the Endangered Species Act (ESA), commonly referred to as Section 10(j). Pursuant to 1982 amendments to the ESA, the agency is allowed (but not required) to designate reintroduced bears as *nonessential* (a population whose loss would not be likely to appreciably reduce the likelihood of the species survival in the wild) and *experimental* (a population wholly isolated from naturally occurring populations of the same species).

The "nonessential experimental" designation gives managers considerable flexibility. As explained in Appendix 12 to the Bitterroot draft EIS, Section 10(j) exempts the reintroduced population from the normal requirements of federal consultation and critical habitat designation and allows the FWS to write special rules "to tailor the reintroduction of an experimental population to specific areas and specific local conditions, including specific opposition."

Section 10(j) does not expressly authorize the creation of a citizen management committee, but such authority reasonably may be read into the secretary's broad regulatory authority to conserve threatened species, together with the encouragement to adapt recovery efforts of experimental populations to meet local concerns.

The coalition presented the plan to the FWS for its consideration as a new alternative. The agency ran it by attorney Margot Zallen at the Interior Solicitor's Office in Denver. As she explained it recently, the key to keeping such a committee within the bounds of the law is to make sure the secretary of the interior retains ultimate decision-making authority. "It's not really delegating authority," she noted, but rather "authorizing" the committee to make certain types of decisions under the secretary's supervision. She made a number of changes, but concluded that the basic proposal was legal.

By July 1997, when the FWS published the draft EIS, the agency's preferred alternative included a detailed proposal for grizzly bear management by citizen management committee—in its key elements, the very plan cooked up by the conservationist-timber coalition.

Alternative One

The draft EIS for Bitterroot grizzly bear reintroduction sets out four alternative approaches. Alternative One, sometimes dubbed the "Citizen Management Alternative," is the proposed action, the alternative currently preferred by the FWS. Its key points are the legal status of the reintroduced grizzlies ("nonessential experimental," as described previously) and their management by local citizens.

If implemented as currently proposed, Alternative One would authorize the secretary of the interior to appoint a fifteen-member Citizen Management Committee (CMC). Committee members would include seven individuals recommended by the governor of Idaho, five recommended by the governor of Montana, one recommended by the Nez Perce Tribe, one representative of the U.S. Forest Service (appointed by the secretary of agriculture), and one representative of the Fish and Wildlife Service. Among the state-recommended members there must be a representative from each state's fish and game agency.

The proposal requires that the CMC members "shall consist of a cross-section of interests reflecting a balance of viewpoints, be selected for their diversity of knowledge and experience in natural resource issues, and for their commitment to collaborative decision making." Moreover, CMC members "shall be selected from communities within and adjacent to the Recovery and Experimental areas"—respectively, the 5,785 square miles of designated wilderness where the bears would be released and the 25,140 square miles of public lands in central Idaho and western Montana within which the bears might be expected to roam.

The CMC would be required to prepare two-year work plans for the secretary of the interior's review, outlining directions for the Bitterroot reintroduction effort. The CMC's most important responsibility is to develop plans and policies to manage the bears in the Experimental area, where grizzly occupancy is to be "accommodated" with human uses.

Above all, the committee would operate under this mandate: "All decisions . . . must lead toward recovery of the grizzly bear and minimize social and economic impacts." If the CMC's actions do not result in recovery, the secretary may disband the committee and resume lead management implementation for the Bitterroot grizzlies. The proposal spells out a six-month-long process by which the secretary and the CMC are to try to resolve differences.

The draft EIS was released in July 1997. The Fish and Wildlife Service hosted a series of public meetings in Idaho and Montana, which demonstrated the great divide in public opinion concerning the bear. Nearly every speaker in Salmon, Idaho, for example, expressed outright opposition to the return of the grizzly. The opposite was true in Missoula, Montana, where most speakers favored full protection for the grizzly and broader habitat designation.

Alternative One, setting forth the idea of the CMC, seemed to be left hanging in the void between these two ends of the opinion spectrum. And, while the CMC component of the proposal has attracted some support in written comments (ranging from the Missoula County Commissioners to the national AFL-CIO), coalition members have generally found their collaborative approach a hard sell.

In light of the contentious public comment period, the Forest Service decided to cast its vote in favor of a time-out. A letter from the regional foresters in the two regions affected by the decision concluded that "the success of this endeavor hinges upon broad public acceptance of a method for bear recovery. The decision-making process should allow more time for an active public outreach program that reaches down to county and local groups, elected officials, and concerned individuals." The regional foresters suggested that the EIS be rewritten to better reflect public concerns and to "fashion a more supportable alternative."

The Radical Center Draws Fire

"Committees are only as good as the people who create them," says Mike Bader, executive director of the Alliance for the Wild Rockies, when asked about the Citizen Management Committee. "What do you think you're going to get with the governors of Idaho and Montana charged with appointing committee members? Each has a track record of making timber industry appointments to sensitive positions like this."

Most people expressing reservations about the CMC repeat this point. How can we trust the management of a sensitive species like the grizzly to a politically appointed committee, especially when its members are drawn from the communities in which opposition to the grizzly is most virulent? (Conversely, one opponent of grizzly reintroduction expressed a fear that the CMC "gives control of land management to the special interest groups supporting the proposal.")

While efforts at sabotaging through hostile appointments are possible,

CMC supporters believe that reason will prevail both in the appointments and the committee's work. "It should be in the best interests of the governors' offices to recommend people who can best support the goals of recovery while minimizing social and economic impacts," points out Dan Johnson. And, adds Hank Fischer, the fact that all parties have so much at stake means that "people serving on the committee will feel that they have a lot of responsibility."

Tom France agrees, pointing out that the timber industry has a lot riding on this proposal, and thus will pressure the governors to appoint people who will help make it succeed. As a practical matter, he does not believe that grizzly bear reintroduction is likely in the foreseeable future without the kind of local support the CMC is designed to build.

The proposed rule appears to allow the secretary of the interior to determine whether recommended committee members meet the criteria listed earlier. Thus, presumably, the secretary could nix any candidate who fails to demonstrate a "commitment to collaborative decision making" or a dedication to the committee's goal of ensuring the grizzly's full recovery in the area. But, other than a provision allowing the secretary to appoint committee members if a governor fails to recommend them, it's not clear how such authority would work.

Beyond the initial question of committee membership, objections to the CMC fall into two categories, representing opposite ends of an ecopolitical spectrum. At one end are those who believe that the committee will work—too well—and will shift essential authority away from the Fish and Wildlife Service. Literature distributed by the Alliance for the Wild Rockies warns that the CMC "will run grizzly bear management, instead of the U.S. Fish and Wildlife Service." Alliance director Mike Bader says that the CMC is "improper because no group of local citizens can serve national interests." People living close to natural resources, he adds, often make the worst decisions about their management.

Bader sees Alternative One as a ruse to loosen restrictions on logging and other development on federal lands throughout the 25,140-square-mile Experimental area. "It's an end run around the courts, Congress, and public opinion," he concludes.

Grizzly bear researcher John Craighead agrees that the CMC proposal is a dangerous precedent. In a *Missoulian* editorial dated August 18, 1997, he queried why the Fish & Wildlife Service "is willing to renege on its congressionally mandated responsibilities and hand them over to nonprofessionals."

Brian Horejsi, a Canadian wildlife scientist and a board member of the Alliance for the Wild Rockies, has nothing good to say about the CMC proposal. In an editorial published in the *Missoulian* on October 16, 1997, he warned that devolution of land management in his country had "paralyzed democracy" by favoring special interests who benefit from resource development. "The Forest Service and FWS are by no means perfect," he wrote, "but compared to management of local citizen committees and governor's decree, they represent the end of the rainbow."

Thus, in short, these objectors fear that the CMC will compromise the relatively straightforward and transparent process by which federal agency officials make decisions about grizzly management. They don't want to lose the ability to influence these decisions at well-defined points—through comments, administrative appeals, or litigation. At the very least, the CMC represents a dilution of authority, and thus a fuzzier target for anyone seeking to challenge its decisions.

On the other hand, many voices have risen in opposition to the plan from those who fear that it represents more of the same—federal control—but in a deceptively shiny new package promising local influence. Idaho's Governor Phil Batt, for example, told the U.S. House Subcommittee on Forests and Forest Health in June 1997 that he's "skeptical of the Fish and Wildlife's sincerity. . . . Make no mistake about it, this plan keeps Washington, D.C., in the driver's seat." Montana's Senator Conrad Burns wrote in a July 1997 *Missoulian* editorial that he regards the CMC proposal as "little more than window dressing. This panel won't have much real power."

A representative of the wise-use group People for the West! stated in September 1997 that he liked the concept of citizen management, but that the secretary's veto authority is "outrageous." Instead, suggested Dave Skinner, the CMC's decisions ought to be subject to review by the governors and secretary every fifteen or twenty years.

For his part, Montana's Governor Racicot has conditioned his support for the plan on several additional provisions aimed at preventing the secretary from using "a pretext to reassume the authority granted to the CMC." Among his suggestions, he recommended that the secretary and committee members submit any differences to a three-member scientific panel (essentially a nonbinding arbitration process) and that the secretary would consult with the governors of Montana and Idaho before making final decisions based on the panel's recommendations.

The Confederated Salish and Kootenai Tribes, which support grizzly reintroduction, also expressed concern about the power of the CMC. Tribal Council Chairwoman Rhonda Swaney wrote in her comments that the tribes "are concerned about the potential lack of effectiveness of such a group if it does not have the ability, both real and legal, to institute its recommendations."

These are divergent views, to be sure. They reflect a complementary pair of distrusts that pervade discussions of public resources: the distrust of rural westerners for most anything preceded by "federal," and that of environmental activists for most anything preceded by "local."

But these apparently divergent viewpoints are also, oddly enough, accurate reflections of the proposal's ambiguities. It does not hand over day-today management authority to local citizens; that would continue to be the responsibility of federal and state resource managers. Conversely, it does limit the ability of the FWS to make policy decisions concerning the grizzly bears and their habitat; while retaining ultimate decision-making authority, the secretary of the interior must cooperate with local citizens.

Thus, in a sense, everyone's fears are valid. The CMC relies on faith. Faith that the committee will be made up of competent people who sincerely want to make it work. Faith that the federal agencies will defer appropriately to the committee's decisions but act decisively when necessary. Faith that the grizzly will benefit from all this cooperation.

And, in the end, that final article of faith may be the trickiest. As Thomas McNamee remarked in a recent telephone conversation, grizzly management is "such a piece of biopolitics." The bears, he said, "are so extraordinarily sensitive to disturbance, and so slow to reproduce . . . their recovery is too precarious to trust to accommodating management."

The flurry of talk has focused on how decisions will be made if and when the grizzly returns to the Bitterroot Mountains: Who will really be in charge? How will the various interested parties be able to influence decisions? And what will it mean for other reintroduced animals? The needs of the bear itself have received far less attention than the potential precedents that the plan might establish.

This is not uncommon for such high-stakes discussions. Collaborative approaches tend to be most promising when applied to solve problems among people with an incentive to develop and maintain relationships with one another. As issues become larger and involve more entrenched interests—the abortion debate and wilderness designation come to mind—the likelihood of cooperative solutions diminishes.

The citizen management proposal appears to be well crafted, solid in its legal authority, and impressive in its detail. It should and probably will serve as a model for other community-based conservation initiatives. If it is adopted, the CMC will be scrutinized at least as closely as the high-profile Quincy Library Group. And, as Quincy has demonstrated, the bright lights of public attention can be harsh indeed. The grizzly bear—that most charismatic of megafauna—may simply be too hot a subject for such innovative experimentation.

"Recovery of the grizzly bear is not a biological problem," FWS grizzly bear coordinator Christopher Servheen remarked to Thomas McNamee, "it's a social problem."

EVALUATING COLLABORATIVE
CONSERVATION:
A CHAUTAUQUA

The collaborative conservation movement underscores some of the deepest political fault lines in the West. For many, the notion of collaboration has gotten caught up in the devolution debate: To what extent should the management of federal lands and resources be delegated to lower levels of government? To others, however, collaboration usually has little or nothing to do with devolution but rather is a new means to implement policy under existing decision-making authorities. Some look to collaboration as an improved way of crafting new policies—ones that may have a better chance of successful implementation. Still others raise the question of measurable results. How can we assess the effects of collaborative decisions? How can we know—or can we know?—if a series of decisions made through collaborative processes are as good as, better than, or worse than decisions made through the more conventional processes?

The following essays are examples gleaned from the emerging, robust debate over the merits of collaborative decision making. The first two essays present a straightforward dialogue between two men who care deeply about environmental politics. Law professor George Coggins argues that collaboration is just the latest fad in a long series of fads that have professed to challenge the existing paradigm of environmental decision making. But, Coggins argues, this too shall pass. The U.S. Constitution predetermines the chain of command in federal resource decision making, and any form of devolution is bound to fail the simplest constitutional tests.

Politics professor Phil Brick shoots back: Give peace a chance. Brick argues that in the vast majority of cases, collaborative efforts do not seek to supplant federal decisions—indeed, collaboration will likely add only slow, incremental changes to existing policies. But in the meantime, collaborative roundtables can build trust, raise general awareness of environmental policy, and enhance the gains of environmentalists by helping them move beyond "green hubris."

David Getches and Douglas Kenney, both of the University of Colorado, comment on the burgeoning phenomenon of watershed groups in the West. Getches asks two provocative questions: Can watershed groups be sustained without an immediate cause to motivate their members (in other words, is "place-care" sufficient motivation to overcome the rigors of endless meetings and the built-in contentiousness of such coalitions)? Second, can watershed groups succeed without strong government involvement?

Kenney addresses the question of effectiveness. How do we measure the effectiveness of watershed groups? Is "political effectiveness" sufficient? If a given group builds trust and enables adversaries to work together, is that any guarantee that real environmental progress will occur? Or might the improved relations be based on the avoidance of tough issues?

Andy Stahl, executive director of Forest Service Employees for Environmental Ethics, argues that ownership and accountability lie at the heart of good stewardship; absent those, government agencies are free to make all manner of irresponsible decisions. Stahl fears that with the passage in 1999 of the Secure Rural Schools and Community Self-Determination Act, Congress has tried to pass the buck in the matter of responsibility over important portions of the U.S. Forest Service budget. There is nothing politicians like more than collaboration, argues Stahl, for delegated decision making often takes elected officials off the hook.

The concluding essay offers an examination of five paradoxes built into the collaborative approach. Using the Applegate Partnership as the setting for his investigation, Jonathan Lange analyzes paradoxes of power, stakeholder involvement, and psychological resistance that seem to emerge organically from a coalition of adversaries. Lange's essay offers a unique perspective from the point of view of a professional educator/facilitator.

Of Californicators, Quislings, and Crazies: Some Perils of Devolved Collaboration

George Cameron Coggins

"Devolution," "collaboration," "community," "dialogue," and "consensus" are the latest buzzwords in federal land management policy circles. They join a list of other undefinable if not undecipherable concepts such as multiple use, holistic management, deep ecology, wise use, biodiversity, and ecosystem management in the pantheon of panaceas. Library groups (Quincy), partnerships (Applegate), and similar forms of gatherings are all the rage at public land law conferences.

The collaboration notion is the raison d'être for numerous new organizations throughout the West, and low-level consensus building apparently has been embraced by an entire spectrum of federal land users and advocates, from tree-huggers to land rapers. The theory (or a theory) evidently is that a self-selected group of people who promise to be civil with one another can do a better job of planning and managing than the duly constituted authorities. Many proponents of collaboration and the like are careful to add a caveat that the collaborative process is not a panacea, but some federal land managers apparently regard it as just that: an all-purpose method for passing the buck on difficult allocation and regulation questions.

Only a misanthropic curmudgeon could question the utility of a negotiating or mediating mechanism that stresses the virtues of civility, reasonableness, cooperation, and consensus building. Oh, well. This essay nevertheless contends that devolved collaboration is not only not a panacea, it often is not

163

even a good idea. I am an outside observer, a law professor without significant experience in federal land management, collaborative or otherwise. My skepticism arises from several decades of researching and writing about public natural resources law and innumerable conversations with the various actors in those legal and political dramas. This essay concludes that the law and its processes, imperfect as they are, are still far preferable to local negotiation as means for resolving public resource issues.

This is not to say that collaboration, consensus, and community are bad. Quite the contrary. Anyone who wishes to collaborate with anyone else concerning their own properties and local community issues is free to do so and should be encouraged to do so. Indeed, more consensus and less litigation likely would make the world a better place. Consensual transactions—The Nature Conservancy is a model—can serve all kinds of private and public values. This country, especially the western regions of it, has never had overmuch civility. When the subject at issue is national lands and resources, however, devolved collaboration is inappropriate for a host of reasons, not the least of which are human history and human nature.

The first part of this essay argues that collaboration as currently envisioned is seriously, if not fatally, flawed. After describing briefly the context of the New West and the new legal standards governing the public lands, the essay elaborates on some of the arguments against collaboration/consensus: Many of the premises underlying them are unproven or untrue; collaboration can be simply a mask for cooptation; collaboration without clear legal direction is ineffective and unenforceable; local collaboration can interfere with important national plans and policies; and collaboration implicates abdication of legal responsibilities. The conclusion argues that the rule of law is superior.

Toto, I Don't Think We're in the Nineteenth Century Anymore

That the American West has changed greatly is a truism. The consequences and manifestations of the changes already undergone and in progress or prospect, however, argue against reliance on local collaboration as the dispute resolution mechanism of choice.

The West has become the most urbanized area of the country. The immigrants to western cities and suburbs frequently reject the values or premises of the old-timers residing in rural areas or corporate boardrooms. Urban dwellers usually look more favorably on opportunities for recreation than they do on maximizing red meat production or subsidizing irrigation. The intermountain West has been the fastest growing area of the country, even

though all of the traditional economic activities—mining, grazing, logging—
have experienced severe declines.

No one anymore believes that livestock grazing on poor-condition public
rangelands is necessary to feed the world or the nation or even Lovelock,
Nevada. The cries for development of "strategic minerals" or "energy independence" of the 1970s so far have turned out to be so much bombast. Predator eradication to provide crops of deer, elk, and other animals is now a universally discredited technique (except possibly in Alaska). The precipitous
drop in timber harvests since the northern spotted owl, marbled murrelet,
and salmon runs were listed as endangered or threatened has had little impact
on the thriving economies of Oregon or of the United States. Recreation in its
many forms has become the dominant economic resource in the West, eclips-
ing all the commodity resources by orders of magnitude.

In the new and far more complex western United States, a few neighbors
gathering to determine the fate and future of the neighborhood or county
simply will not work—if it ever did. The Old Westerners resent one-man-one-
vote democracy because there are fewer of them and more of the interloping
Californicators. Voluntary enlistment in a collaboration venture cannot assure
representation of all who have a legitimate voice (including eastern tourists)
nor consideration of all legitimate interests (including future generations).

The nature of public natural resource controversies has changed irrevoca-
bly in the past several decades, and even bigger changes seem nearly inevitable.
The legal focus has shifted from private disputes to larger public questions.
The important question no longer is whether, for instance, Acme or Mammoth gets the federal oil and gas lease, but rather whether any petroleum
development should occur and, if so, under what protective conditions.

This is so for several reasons. The "public interest" groups, which do not
have financial stakes in litigation or lobbying outcomes, have grown exponentially since the 1960s while the agricultural and extractive resource industries and the numbers of their employees have declined, absolutely as well as
relatively. The environmental organizations have become the chief enforcers
of some environmental statutes, notably the Endangered Species Act and the
National Forest Management Act. Those statutes proliferated in the 1970s,
and their main thrust is preservationist.

The courts have opened their doors to this new form of litigation—somewhat grudgingly—and Congress has endorsed this trend by enacting dozens
of citizens' suit provisions in environmental legislation. Litigation first in the
federal agencies and then in the federal courts has become a standard part of
the federal resource allocation process, and only the most radical legislation
or judicial decisions could alter that fundamental fact. We have always been a
litigious people, and public interest litigation fits, though somewhat uncom-
fortably, in the national mold.

Cowboys: Check Your Assumptions at the Door

The general reasons for skepticism about reliance on enhanced consciousness are buttressed by a host of more specific reasons. I start with the proposition that the premises on which the collaborationists rely are false or unproven.

The first assumption underlying most proposals to devolve authority onto local collaborationists is that, at bottom, we are all reasonable people who will see both sides of the issue and reach appropriate compromises. This assumption is demonstrably untrue.

In the West today may be found people who insist that federal land ownership is unconstitutional as well as immoral; who shoot, shovel, and shut up; who exploit to get theirs before the Second Coming; who spike trees; who bomb federal offices; who threaten and kill federal employees; who routinely commit perjury; who monkeywrench; who steal government property; who blame NAFTA on Jewish bankers; who are entirely dependent on forms of federal welfare; who are New Age hippie mystics; and who cordially despise each other.

The West is the home of the worst demagogues (e.g., Steve Symms) and demagoguettes (e.g., Helen Chenoweth) in the country. No one can claim that all or even most westerners are innately unreasonable, but the observable number of whiners, crooks, blusterers, and outright crazies in western states seems out of proportion to population numbers.

A second main assumption of the collaborationists is that agreement or consensus is possible. In fact, however, many issues in natural resources law and management are not susceptible to unanimous agreement or even general consensus.

Montana ranchers and Defenders of Wildlife, for instance, will never reach common answers to such questions as whether wolves should be reintroduced to Yellowstone or whether bison migrating out of Yellowstone should be shot. The Sierra Club, whose members have voted to seek a total ban of logging in national forests, cannot conceivably reach any real agreements with the logging industry or the Forest Service, who are committed to the allowable cut. Senator Hatch and the Southern Utah Wilderness Alliance will never agree on the appropriate extent of official wilderness.

No person, group, or industry in the West will give up its historical preferences, subsidies, and special treatments without a to-the-death fight in Congress and courts, no matter what. The more extreme elements of the phenomenon called the environmental movement—Greenpeace, Earth First!, Sea Shepherd, monkeywrench gangs et al.—cannot loosen their devotion to radical remedies without deserting their identities.

The point is that the values and premises of participants in resource allocation controversies sometimes—perhaps most times—differ irreconcilably, and no amount of talking is going to change that.

The third main collaboration assumption is that negotiated plans and agreements will leave all participants and the public interest better off. Not necessarily so. Every such controversy is going to produce losers as well as winners. Collaboration is seen as necessary precisely because some variety of resource damage or shortage is inevitable.

Some of the participants will be the same people or entities that have wreaked the environmental damage that collaborationists are now trying to undo. Given past overexploitation of grass, timber, water, and wildlife, we are no longer playing a zero-sum game; economically, public resource allocation in the near future necessarily will be a minus-sum game.

To argue that an agreed, optimum solution will leave all players better off is sheer sentimentality. Distributing the losses is a more likely outcome.

A fourth main assumption is that the issues on which collaboration is needed are local problems better solved by local people with knowledge of local conditions. In a few cases, this may well be true. For the most part, however, allocation and conservation of federal lands and resources necessarily and properly are national, not local, questions.

The Constitution entrusts the disposition, regulation, and care of federally owned assets to the national government, not to local self-appointed mediators. Because every significant question involving allocation of public resources ultimately is a value question—which is to say, a political question—it should be resolved by the legitimate political process.

Collaboration as envisioned by the visionaries is merely politics by another means, but the means are not those created by the Constitution or endorsed by Congress, the responsible political body.

A fifth implicit collaboration assumption is that the federal government is the bad guy. Citizens must take these matters into their own hands because the federal government is (choose one or more) corrupt, inefficient, bureaucratic, insensitive, overexploitationist, underexploitationist, democratic, ungodly, inept, unnecessary, or, simply, The Enemy. None of these charges is indisputably false, but most are unproven, and all are irrelevant.

Whatever the faults of the Bureau of Land Management or the Forest Service, the appropriate remedy is neither abdication nor vigilantism, both attributes of collaborationism. The federal government is the only national government we have. It owns the federal lands and resources, and it must be responsible for allocating them in the fashion that a national majority—not a local group or partnership—deems appropriate.

In sum, at least some of the assumptions on which the case for collaboration is premised are contrary to human nature and human experience. Our legal system and our history recognize the inevitability of disputes and controversies not amenable to resolution by consensus building, and we have established an intentionally inefficient tripartite representative machine to resolve those disputes without bloodshed through law.

Warning: Collaborative Community Dialogue Can Be Hazardous to Your Principles

Some environmentalists suspect that the wave of enthusiasm for devolution/collaboration springs more from a desire to defend the West's peculiar caste systems from the onslaught of modern reality than it is a good faith effort to work out mutual problems. National conservation organizations especially are leery of devolving decision-making powers onto those who stand to benefit the most.

History supports the surmise that collaborationism in at least some cases is an insincere rear-guard holding action. Certainly, ranchers traditionally have had no desire to include nonranchers in advisory councils, and timber companies had no apparent inclination to broaden forest decision making beyond public and private foresters. The mining industries have consistently opposed and still resent legal mechanisms allowing ordinary citizens to have a say in the desirability of mining operations. Professional game and fish managers long have regarded nonhunters and antihunters as the ignorant unwashed.

The industry and agency passion to find consensus evidently coincides with the degree to which the noneconomic interests have been successful in asserting newer and broader management priorities.

It is thus at least conceivable that the big but declining economic interests in the West—the large irrigators; the large ranches with federal privileges; the major oil, coal, and hardrock mining combines; the entrenched concessionaires—wish to use devolution and collaboration as means of cooptation to forestall the far more radical reforms looming on the legal horizon. All of these interests may try to hide behind the so-called little guys (small ranchers, miners, concessionaires, etc.), but they have the larger vested interest in the status quo due to the subsidies and privileges that permeate public resource allocation.

I do not pretend to know the true motives of collaboration participants. But the all too frequent reference of the participants to vague concepts such as "community" and "lifestyle" indicate that collaboration is often intended as a means of status quo preservation. These "custom and culture" notions have already engendered the silly and sometimes destructive county supremacy movement. Similarly, the oft-repeated Chicken Little alarms about looming bankruptcies and other catastrophes should change occur buttress the surmise that consensus building is a cover for prolongation of unjustifiable subsidies, preferences, and privileges. The whole idea seems somewhat redolent of James Watt's Good Neighbor policy, a cover for wholesale ripoffs of the U.S. Treasury.

What House of Cards?

Groups assembled to collaborate are transient entities. Without guaranteed long-term funding, they will have no professional staff, and thus will lack an institutional structure. Without a permanent institutional structure, they are destined to wither away as the perceived crisis passes. Voluntary collaborators (unlike other well-known collaborators such as Quisling and Petain) have no coercive enforcement powers, so parties may depart from collaborative solutions with relative impunity. Outsiders, of course, do not have even an ethical duty to go along.

When the groups dissolve, as they will, the entrenched economic powers will remain, and the old status quo will reemerge. New abuses then will cause the process to start all over. Without the force of law, local plans and local allocative solutions are written on the sand.

Because collaborative agreements lack coercive powers, they also will be less efficient in conventional terms. The inefficiency of (and bad feelings engendered by) litigation is a common reason given in favor of collaborative processes. But unless the consensus solution is embodied in federal legislation, it will have no preclusive effect on any disaffected party seeking judicial review. In other words, if the collaborative effort does not go into the standard lawmaking machine, it becomes only another step in an even more inefficient (and undemocratic) decision-making means.

These Lands are Your Lands . . . Not

The federal lands are national assets, not local fiefdoms. The public natural resources belong to the entire American public; they are not just local storehouses to be looted in the deregulation riots. Local collaborative solutions could impede or contravene national laws or policies.

Congress in its wisdom has decreed that allocation of national assets must be preceded by various procedures and must be in accordance with certain substantive standards. The Administrative Procedure Act requires federal agencies to observe certain procedures, such as notice-and-comment rulemaking, that local groups might well be tempted to circumvent. Similarly, the formal consultations called for by the National Environmental Policy Act and the Endangered Species Act could easily be seen by local interests as contrary to their desires even though embedded in federal statute.

Even more to the point, local collaborators will have little if any incentive to ensure adherence to positive, substantive legal rules. For instance, local participants could easily arrive at a consensus in favor of an economically valuable project that requires dredging and filling of wetlands, while the EPA, with a national view of overall wetlands decimation, might take a considerably more jaundiced view.

As a strictly legal matter, there is no question but that national law overrides contrary local decisions. As a practical matter, however, if consensus mechanisms were accorded high status, the national goals would be far more difficult to achieve.

Recent years have seen growing sentiment for larger-scale management priorities. The concept of a Greater Yellowstone Ecosystem is gradually coming to fruition. The Poppers' idea of a Buffalo Commons on the High Plains has no visible political support but makes considerable sense nevertheless. Large-scale regional plans (Pine Barrens in New Jersey, Tahoe in California/Nevada, the Columbia Gorge in Washington/Oregon) are growing in popularity and complexity.

Land-use planning for federal land tracts must embrace more than local concerns. The fight to save endangered and threatened salmon runs in the Pacific Northwest necessarily implicates multiple causes, multiple jurisdictions, multiple agencies, multiple parties, and multiple remedies: Local collaboration can never effectively deal with a problem of this magnitude. Some recently have advocated creation of a wildlife supercorridor running from Yellowstone to beyond Lake Louise in Canada. Others are promoting a National Heritage Trust to increase and consolidate preservationist land holdings. These are national proposals of national importance and magnitude. Local collaboration would only sidetrack national solutions to national issues.

Let's Not Kill All the Lawyers

New collaborative mechanisms and structures and methods are unnecessary. All of these efforts try to work outside legally ordained decision-making processes. Those processes are far from perfect, but they are far superior to vigilante circumvention of them. To iterate a few basic lessons:

Ours is a representative government in which the president and the Congress are elected by the people, and the judiciary and high-level bureaucrats are appointed by the executive and confirmed by the legislature. The Constitution gives the Congress the power to dispose of and make "needful" rules for federally owned property. Congress has chosen to delegate much of its authority to federal land management agencies but has restricted their discretion by enactment of dozens of laws dictating or indicating outcomes in certain situations. Congress has also instructed the agencies to consider the environmental and other consequences of their allocative decisions, and, in the process, to consider (often in writing) the views of all interested governments, citizens, and entities.

"Flexibility" has been more of a curse than a benefit in public land management. All too often, "flexibility" or "discretion" merely means deciding essentially the same question differently, usually in response to local political pressure. The public land management laws now on the books already accord managers a latitude that would be considered overbroad in other areas of governmental regulation. Devolution/collaboration increases undue discretion by giving the manager an easy out. The country and the collaboration participants both would be served far better by clear legislative and regulatory rules that confine administrative discretion, not expand it.

Current federal law does not recognize collaboration/devolution as an appropriate means of public land decision making. The statutes typically invite input from the entire spectrum of those concerned and demand that decisions be made openly and publicly, but nowhere do they delegate decision-making power to unelected, unappointed citizens at large or interested economic entities. Agencies, not ad hoc groups, are to allocate resources according to the (all-too-often-negligible) congressional guidelines. Losers have the options of appealing to courts and Congress. This system, by and large, works. It is inefficient, but inefficiency is built into a tripartite government that requires certain democratic procedures as well as certain substantive results.

The flaws in the current legal system will not and cannot be cured by the New Age wishful thinking that believes all problems can be solved by just sitting down and talking. The lamb may lie down with the lion, but it will not get much sleep. Some form of coercion is absolutely necessary, and it can come only from the law. Ours, after all, is supposed to be a government of laws, not of men.

As I have written before, in a slightly different context, "Instead of allowing federal land managers to devolve their authorities and responsibilities onto local citizens' councils, a far better balance will be achieved if only legislators would legislate, judges would judge, and managers would manage in accordance with law."

Of Impostors, Optimists, and Kings: Finding a Political Niche for Collaborative Conservation

Philip Brick

Although George Cameron Coggins generously describes himself as a misan thropic curmudgeon, his critique of collaborative conservation betrays a quixotic if not naive optimism. After accusing community conservationists of "New Age wishful thinking that believes all problems can be solved by just sitting down and talking," Coggins then suggests that "a far better balance will be achieved if only legislators would legislate, judges would judge, and managers would manage in accordance with law." With the salvage rider and the failure of mining and range reform in recent memory, is this not also an extraordinary exercise in wishful thinking? Can we really expect the system to work so well without us?

Because some of my best friends are misanthropic curmudgeons, and I secretly aspire to such status in my old age, I almost feel compelled to defend the honor of curmudgeons everywhere by shouting, "Coggins is an impostor! Seize that man and send him to curmudgeon therapy!" (Surely, there must be a twelve-step program for wayward misanthropes somewhere.) But alas, there is much in Coggins's essay to agree with. He is surely right to argue that community-based conservation is no panacea, and that issues of national importance ought to be resolved through national, not local, processes. But just because local collaborative conservation efforts can't solve all our problems does not mean that they can't begin to address some of them. Similarly, just because reliance on national activism and law is no panacea does not imply that it is not indispensable.

Local collaborative conservation efforts are not only a good idea, they need not be exclusive of efforts by national environmental groups to lobby and litigate at the national level, as Coggins suggests. The emergence of library groups and partnerships, if properly understood, can strengthen the environmental movement instead of dividing it. Much of the conflict between local conservationists and their national counterparts is the result of misperceptions and misplaced aspirations on both sides.

Local partnerships represent an innovative approach to public lands management, and as such appear to threaten existing policy structures. Some local conservationists, frustrated with policy gridlock and the skepticism of national groups, have done little to dispel this notion. Indeed, if I spent many unpaid nights and years at a local library table, hammering out solutions with people I disagree with, I too would like to think that the implications of my work would be "revolutionary."

But if the history of public lands management tells us anything, it is that existing policy regimes, embedded in national environmental laws, bureaucracies, and political institutions, are incredibly durable and resistant to change. Local partnerships therefore must do a better job explaining how their work can be integrated into these national policy regimes.

Without such an explanation, it is only natural for national environmental groups to be wary of local collaborative processes that appear to threaten the power and unity of the environmental voice in public land policy debates, or worse, threaten to fundamentally transform existing policy regimes at the very moment environmentalists appear to have more than just a foot in the door. But despite their differences, local conservationists and national environmentalists suffer from the same illusion: that public lands policy regimes are amenable to rapid change and can be reshaped to serve a finite and exclusive set of policy preferences. The future of public lands management is likely to be much more mundane—it will continue to be characterized by incremental modifications to existing, national policy regimes, but hopefully with a flexibility and creativity that can come only from experimentation on the periphery—from local partnerships.

Although I am a supporter of local collaborative efforts, I continue to believe that, at least for the time being, national management of our public lands offers the best hedge against the unfettered hegemony of economic interests, which can often command local majorities but not national constituencies. The environmental movement is well equipped to lobby and litigate at the national level, winning a few small battles and forcing gridlock where it can't win. But the environmental movement is also mired in its own Vietnam—winning many inconclusive battles but losing the war.

Community-based conservation can help, but it must be conceptualized in a way that strengthens environmental objectives instead of weakening them. For example, community-based conservation should not get into the business of "logging without laws." That can be left to Congress, which Coggins (without a trace of irony!) calls "the responsible political body." But where innovative, local plans meet or exceed national standards, environmentalists should be more concerned about results and less worried about precedents that may appear to weaken existing policy regimes.

A Political Niche for Collaborative Conservation

In America, it seems that simply having a good idea is never enough. Instead, we have to apply what might work in one place everywhere. Thus we have model schools, model welfare programs, and model watersheds. Although much of this hype is the predictable product of competition to fund local social and political experiments, we are far too prone to believe in panaceas. From their infancy, local partnerships have been held up as models for the future of natural resource policy management. Instead of gridlock, we can find consensus and move forward. But no one has really been able to articulate where that "forward" might be. Are we really ready to devolve the management of public lands to local consensus councils? How are interests not at the local table to be represented? How can scientific expertise be integrated into citizen planning and management? Can the power and resources of industry be matched at the local level? These are indeed vexing questions, though we should not, as Coggins suggests, reject local collaboration just because it can neither resolve these questions nor offer a convincing alternative to federal land management.

Instead of expecting local experiments at collaborative problem solving to evolve into radical new directions in public lands management, we might instead appreciate what collaborative conservation might contribute to existing arrangements. I see four important functions.

1. Advise and Innovate. Bureaucracies are not known for finding creative solutions to complex social and political problems—they simply are not set up to do this, despite the "can do," technocratic-utilitarian ideology of agencies such as the Forest Service. Most public land management problems are more political than scientific, but political problem solving is beyond the traditional and legal mandate of the agencies. Local collaborative groups can provide valuable information and "signposts" for both land managers and nonlocal interests, offering a more accurate gauge of what is politically possible, not just process expedient. Local partnerships may be the best tool we have to find solutions to intractable political disputes in an agency environment that is increasingly driven by procedures, not results. These solutions, in turn, would still have to withstand national scrutiny and conform to national laws.

So while I disagree with Coggins on the intrinsic merit of local partnerships, I share some of his concerns about any attempt to institutionalize local collaborative processes, albeit for different reasons.

Local collaborative groups can be innovative only to the extent that they remain non-institutional. Attempts by agencies to convene partnerships, though well intentioned, have been unable to match the creative energy of more free-forming advisory groups, and agency partnerships have run into thorny legal problems.

2. Build Social and Political Capital for Environmental Goals. In international conservation efforts, the need to work with local communities and indigenous peoples is axiomatic. But in this country, the environmental movement often ignores this wisdom, preferring instead the hammer of national environmental legislation to accomplish its goals. It is no wonder that opinion polls in rural areas show steady support for environmental concerns, but increasing contempt for "environmentalists."

National environmental regulations can compel change, but these will be shallow and short-lived without a corresponding development of local social and political capital. The history of civil rights in this country is instructive. Although no one can discount the importance of national civil rights legislation in ending blatant discrimination, real progress in achieving equality is more a function of local social capacity and consent.

No law can end bigoted resistance—equality has been possible only where it has been translated into the local language and integrated into local practices. Collaborative conservation groups are an important forum where such translation can take place. Some national environmental groups have begun to appreciate this at more than just a rhetorical level. In his January 1998 report to members, president and CEO of The Nature Conservancy John Sawhill writes, "I believe that community-based conservation will emerge as the primary vehicle through which the Conservancy delivers our conservation product. . . . We boast today of being a multi-local organization, but the future will find us even more decentralized, even more responsive to the distinct conservation needs of local communities."

3. Expand Environmental Justice. In recent years, the environmental movement has been roundly criticized for failing to recognize the social justice implications of its activism. Most of this attention has focused on toxics and the siting of hazardous wastes in poor, minority communities. Here, the environmental movement has acknowledged the problem, and has taken steps to address it. But in public lands controversies, the environmental movement has done more to deny the problem than to address it, at tremendous cost to local communities and to the movement itself.

During the timber crisis in the Pacific Northwest, environmentalists naively tried to play the class card in timber-dependent communities by suggesting that the timber industry has no concern for local jobs because it exports raw logs and may cut-and-run at its convenience. This argument went nowhere, perhaps because locals had no trouble figuring out that environmentalists had neither a history of concern for working families in rural communities nor any tangible plan to address their plight.

Now, more than ever, the environmental movement needs working groups who can demonstrate in creative ways that environmental preservation need not conflict with living-wage jobs in rural communities.

4. *Save the Environmental Movement from Itself.* Political power can easily breed hubris and self-deception, but the simple whiff of power can be just as destructive. Having flexed its lobbying muscle to reverse the antienvironmental agenda of the 104th Congress, and having demonstrated its prowess in the courts to virtually shut down timber sales on national forests, many environmentalists now smell blood and seem convinced that it won't be long before it is their turn to be king. The Sierra Club, for example, was for years content to accept widespread logging on public forests while selectively appealing specific sales that threatened scenic values. Now the Sierra Club advocates a total ban on logging in national forests. In this context, local collaborative efforts that include industry interests seem like unwelcome guests at the victory party. Indeed, Coggins's argument that natural resource industries "wish to use devolution and collaboration as a means of cooptation to forestall far more radical reforms looming on the legal horizon" rests on the assumption that such reforms are foregone conclusions.

But this assumption rests on a remarkably shaky foundation. The past is often a better guide to the future than we would like to believe. History shows that policy regimes embedded in the institutions, norms, and practices of the state are notoriously difficult to dislodge, as Clinton and Babbitt learned in their early attempts at grazing reform. In this case, a mere twenty-three thousand permittees were able to resist the president and his cabinet, a sympathetic (but not filibusterproof) Congress, and the full weight of national environmental groups. One step forward, two steps back, better characterizes change in federal land management. Optimists might suggest two steps forward with only one retreat. Only fools assume progress will be unilinear.

There is every reason to believe that environmentalism's traditional opponents will continue translating their economic power into political influence. On this point, Coggins might read his own writing: "No person, group, or industry in the West will give up its historical preferences, subsidies, and special treatments without a to-the-death fight in Congress and courts, no matter what."

National environmentalists would also do well to remember that their approach is not without strategic weaknesses. National environmental groups have large but relatively uncommitted memberships, relying on "checkbook activism" of predominately white, urban, upper-middle-class voting blocs. This invites a host of problems, not the least of which is the bitter and well-organized resistance of working men and women in rural communities. Although Coggins may view their "wise use" protests as the last gasp of an "Old West" that must inevitably give way to the New, it is clear that their concerns will continue to generate considerable political resonance and capacity to thwart change. Community conservation is an important tool to temper some of those voices, which often acquire their zeal because the environmental movement has no mechanism where local residents can simultaneously express anxiety about their livelihoods as well as their concern for the local environment.

Exercising power is not the same as generating consent, and in the absence of the latter, the environmental movement will always find its goals just beyond reach. There is no such thing as a politically competent majority that does not at least take steps to protect the interests of the minority.

The environmental movement would be better off consolidating and "bulletproofing" existing gains against future reversals rather than grasping at unlikely panaceas of power. The story of the Quincy Library Group is instructive here. In the mid-1980s, local and national environmental groups were on the ropes on the timber issue in the northern Sierra Nevada. Desperate to stop rampant logging of old growth, local and national environmentalists developed a proposal that would have ended logging in ancient forests and near streams, but would have allowed continued cutting in roaded, second-growth areas. The Forest Service considered, but rejected the plan in favor of business as usual. But ten years later, the political tables turned, and essentially the same plan is favored by industry and many local environmentalists, but rejected as a "sellout" by some of the same national environmental groups who helped develop the proposal in the first place.

If local collaborative processes are "hazardous" to one's environmental principles, as Coggins suggests, what might those principles be? The only principle I can identify here is the raw exploitation of power.

The Power of the Powerless

If neither environmentalists nor local collaborators can be king, where does this leave us? I am sure that I have made few friends in writing this essay. National environmentalists will surely resist my accusations of political naivete and admonitions to find virtue in local collaborative efforts. Local conservationists will no doubt resent my suggestion that their role remains, at least for the time being, on the periphery of power.

It is a serious mistake, however, to believe that it is always necessary to actually be pulling the levers of power to effect dramatic changes in the system. Playwright, former dissident, and now president of the Czech Republic, Václav Havel, shows us with characteristic grace how the powerless can have real power. Havel tells a parable of a greengrocer in his home country during the communist regime. Whenever the grocer gets a shipment of vegetables, he also gets a poster with slogans such as "Workers of the world, unite!" and is expected to hang it in his store window. He does so without thinking much about the content of the message, since he senses at some level that there might be trouble if he actively refused to display the sign. Remarkably, no one ever stops to read these signs—they are everywhere and part of the "socialist panorama" that reminds everyone that the socialist system in unlikely to change.

The problem, of course, is that the system is built on lies. In Havel's words, "government by bureaucracy is called popular government, the working class is enslaved in the name of the working class, the complete degradation of the individual is presented as his or her ultimate liberation, the arbitrary abuse of power is called observing the legal code, and farcical elections become the highest form of democracy." If this sounds familiar, it's because Havel speaks to us. He says that great political changes do not come with great events like the Velvet Revolution in 1989. Change comes from small acts of resistance, what he calls "living within the truth."

Once the greengrocer can no longer participate in a system that he knows to be a lie, he begins living within the truth. He no longer puts up the sign—a small act of resistance, but one with powerful consequences. Without the sign, the panorama changes. It's not that anyone notices the greengrocer's sign missing from the window. But as others also "live within the truth," people do come to "know" that change is coming, without giving it much thought. The groundwork is thus laid for a future transformation that everyone, even the Communist Party, is ready for and can accept. So from this perspective, great transformations are more the result of changing public sensibility than a changing of the guard.

Integrating collaborative conservation into national environmental strategy is our best hope of "living within the truth." It is a small step in changing the panorama of natural resource management. It is not an act that fundamentally transforms existing land management policy regimes, but is it too optimistic to suggest that new ideas, generated through a process of consensus building from the bottom up, won't be necessary to someday create new regimes that reconcile our differences—urban and rural, economic and environmental, rich and poor—in more productive ways?

Some Irreverent Questions about Watershed-Based Efforts

David H. Getches

Last year I spoke at a conference in France about the burgeoning watershed movement in the United States. The story to be told was not about how appropriate a natural drainage is for resource management. The watershed idea is hardly new to Europe; France has long managed water according to river basins. Although we westerners usually associate John Wesley Powell with his unheeded advice to organize according to hydrologic regions for political decision making, it was a Frenchman, Joseph Nicollet, who suggested in 1843 when surveying the upper Mississippi River basin, that the boundaries of new territories such as Iowa should follow river basin boundaries.

What is new in the United States is the extent and the suddenness of the current American embrace of the watershed ideal. Beyond describing the new watershed movement in the United States for a European audience, my task was to analyze what makes these efforts tick (or not tick).

The message I wanted to bring from this country was not simply that we had discovered the relationship between geography and natural resources, but that by convening diverse interests—sometimes former rivals—we were finding better solutions to natural resource problems at the grassroots than governments were able to develop. Community-based natural resource problem solving indeed has been adopted enthusiastically in watersheds around the country, and I am a booster of the movement. I had some examples in mind and intended to brag about them to make my point. But if the observations about the watershed movement were to be more than romanticizing, I needed to look critically at what was going on and suggest what makes watershed efforts succeed. This required studying what was actually happening in several watersheds and synthesizing criteria from their experiences.

Aided by the work of others, I attempted to extract some ingredients of success for watershed efforts. The 1996 *Watershed Sourcebook* by the Natural Resources Law Center at the University of Colorado, a study in which I had a small part, compiled information on almost eighty recent watershed efforts. I reviewed all the examples in the book, then focused on five illustrative watershed collaboration projects, from the showcase example of the Henry's Fork Watershed Council in eastern Idaho to initiatives in Oregon's McKenzie River watershed, the upper Clark Fork basin in Montana, the Verde River in Arizona, and Clear Creek in Idaho.

My survey enabled me to document a paper with some illustrative examples but also left me with some irreverent questions—questions that one does not dare ask without risking being branded a hopeless curmudgeon like George Coggins. He has opined that such collaborative efforts range from "insincere rear-guard holding actions" to an abdication of government responsibility (see Coggins's essay on page 163). Unlike my colleague Coggins, who dismisses these efforts as "New Age wishful thinking," I am pretty optimistic about the potential for collaboration at the watershed level and I want these local efforts to work. At the risk of eroding the mythology of the movement, however, I pose two questions that could disappoint or offend readers. Coggins will be proud.

Offense may be a necessary by-product of taking a critical, even realistic look at local collaborative efforts. Criticism could chill the enthusiasm of fledgling groups, and it runs the risk of being based on incomplete information. But perhaps asking irreverent questions, even associating them with specific groups, may provoke some constructive debate. I confess that my observations are based on a most unscientific method of analysis; the evidence is largely anecdotal. Therefore, my views should be taken with a dose of salt. At the very least, however, any controversy they spark should support the idea that more careful, systematic analysis of the U.S. watershed experience of the 1990s is warranted in order to determine what is working and why and what is not working and why not.

After looking hard at watershed efforts all over the western United States, I wondered: Can they survive on collaboration alone or do they need a specific cause? Can they be effective without the blessing of government?

Can Watershed Efforts Be Sustained without a "Cause"?

No one can doubt that litigation, demonstrations, and tough lobbying have advanced the environmental movement just as they did the civil rights and feminist movements. Yet confrontation and advocacy have gobbled up resources, alienated people, and sometimes led nowhere. Environmental "wins" produce losers whose political or economic power or perseverance can allow the issue to be reopened. So we have gravitated to collaborative processes to deal with environmental problems out of a sense that disparate and sometimes antagonistic interests can find common ground and mutually beneficial practical solutions to natural resources problems. Ideally, if communities create their own solutions to natural resources issues, there may be no losers and the solution stands a chance of lasting.

When people put aside notions of winning and losing, it is harder to articulate what they are about. Polluters and fishermen, developers and environmentalists, farmers and urban planners may all lay down their arms and agree to sit around the same table. They are each motivated by different and sometimes opposing reasons—protecting fish populations, getting permission to build a subdivision, and the like. But successful collaboration means that they need to move beyond their very specific goals and find a common cause—for example, restoring a section of the river to its former state. This cause has to be something people can believe in and can work together to accomplish. I have heard people say, when asked the group's purpose: "Now we are all talking to one another and meeting regularly; that's enough of an accomplishment in itself."

But I do not believe that the group can last for long unless they have something real to do, something that makes a difference on the ground. Simply getting people to sit around the same table can be difficult and often is the first major step for collaborative efforts. Yet I doubt that the glow of newly found fellowship among former adversaries will last long unless they have a common cause or common threat that puts them to work on tangible goals. So, I conclude that the first element of successful watershed groups is that they must focus on a clear objective, something more than the touchy-feely notion of togetherness.

Now, it seems that some groups have occupied themselves with gathering and distributing information, publishing newsletters, and the like. Maybe this will last, but I would not devote much of my own time to that kind of program unless it was linked with projects or plans that would change the way things are done on the ground. Compare the activities of a few groups:

- In Idaho, the Henry's Fork Watershed Council makes recommendations on development and improvement projects that may proceed or not or may be modified based on what the varied membership of the council says. That's the real work of governance; it matters.
- In the Verde River watershed in Arizona, flows are depleted by groundwater pumping and water quality, and riparian areas are degraded by sand and gravel mining, recreational, and grazing uses. The Verde Watershed Associ-

ation has sponsored an inventory of recreational uses and a watershed assessment, held several seminars, and published a newsletter. It seems to have viewed its educational activities as an end rather than as a platform for further action. The association does not take positions on issues and has no action agenda, product, or problem to solve, so it has not taken responsibility for making tough choices about how the basin's natural resources should be managed. Some basin residents became disenchanted with the group's intentional neutrality and lack of involvement in watershed management and therefore withdrew. The association appears to have restricted itself to accomplishing too little.

- Oregon's McKenzie Watershed Council, like the Verde, is strong on data development but the on-the-ground nexus is greater. The area has experienced diverse resource uses (timber, hydropower, recreation, agriculture, water supply) and has suffered from poor land- and water-use practices in some cases. Because the council effort was initiated when hydroelectric dam owners sought a renewed federal license, there was a sense that planning and management programs would be reflected in changed dam operations. And the council's activities have gone beyond that goal. A GIS database system and water quality monitoring program support planning efforts and specific restoration projects. They have completed an action plan for water quality and fish and wildlife habitat.
- The Upper Clark Fork Council, in Montana, was formed after water development interests became locked in lengthy conflicts with fish, wildlife, and recreation interests. Facilitated by the Northern Lights Institute, the council began by spending many months learning, studying, and educating. Resources were spent on newsletters and surveys, and countless hours spent on meeting and talking. But all of this was a base for actual projects—a phosphate detergent ban, reuse of municipal wastewater for irrigation, and a plan for water resource management that was adopted by the legislature. The council also drafted legislation closing the basin to new water rights and helped to implement a pilot instream flow leasing program. Real stuff.

Can Watershed Groups Be Effective without Significant Government Involvement?

There is a myth that the watershed movement consists of spontaneous, "bottom-up," local efforts that find alternatives to the rigidity of intransigent bureaucracies and one-size-fits-all federal solutions. This myth appeals to populists, to admirers of civic virtue, and to antigovernment westerners. It is satisfying to think of community-based problem solving in terms of locals elbowing out in front of the bureaucrats and taking control of their own situation.

My review of several groups corroborated that they certainly do incorporate more and varied interests and they put more local people to work on the problems of their region than has been typical of approaches generated by distant governments (that is, state and federal governments as opposed to tribes, cities, and counties). This results in the development of creative solutions that are more flexible than federal or state agency prescriptions. Local people are rightly proud of their efforts and more accepting of homegrown solutions that demand money and time or limits on development than they would be if government regulations demanded the same of them. In part, this is because they perceive wider benefits. And in part, it is because they have helped design the program.

But virtually all of the groups I looked at that were doing anything of importance had considerable involvement of government entities, including tribes, cities, counties, and especially the federal government. A closer look at their roles suggests that the government role is not only helpful but critical—a nearly universal element of all those groups that seemed to be working successfully. Some had federal money and some relied heavily on federal information and data. Some met in federal offices or had a federal employee doing coordination. There were federal officials at nearly all tables, on technical committees, on coordinating committees, and sometimes on boards. Henry's Fork Foundation leader Jan Brown said that after the local environmentalists and water users had reached their truce in the Henry's Fork watershed, their challenge was to convince federal officials that "we want you on the bus, but not driving it." And this may best describe what appears to be the essential role of government in the watershed movement.

Government participants bring a variety of essentials to the table—technical expertise, data, staffing, funding. They can provide support, continuity, and credibility. With regard to the McKenzie watershed, the group was locally initiated when the county and the water and electric board became concerned about watershed management, but state and federal governments have been important contributors to its success. Although only five of the twenty members of the council represent state (two members) or federal (three members) agencies, they play a crucial role. The Forest Service, Bureau of Land Management, and Army Corps of Engineers are heavily involved, committing hundreds of hours of staff time each year. The council received start-up funds from several federal agencies and federal appropriations of more than $1 million through the Environmental Protection Agency and Natural Resources Conservation Service. Further, the U.S. Geological Survey (under a contract) is inventorying and reviewing water quality data.

Government agencies cannot dedicate large amounts of staff time and government resources unless they have made a commitment to the watershed process at a high level. Even more important, agency leadership is needed if the solutions developed and proposed by the groups are to be implemented. Plainly, much of the work of watershed groups would be frustrated and their efforts would be for naught if agencies ignored them or refused to incorporate the plans generated by the groups. If the agencies were not flexible enough to work with the groups to refine their homegrown solutions to fit the requirements of the Clean Water Act, the Endangered Species Act, and other regulatory programs, there would be little point for people to convene and meet day after day and night after night to craft their own programs to comply with or avoid the necessity to enforce regulations.

It is remarkable that federal agencies and some states have put such a high priority on participation in collaborative processes. Federal agencies and a few states have embraced the watershed collaboration ideal. They have reasons not to do so. This is an era of lean resources for governments, yet with no legislative mandate, we see agencies spending the countless hours enduring meetings, educating lay participants, and patiently tolerating drawn-out processes to design plans for saving endangered fish or mitigating the loss of wetlands. One might think that overworked officials with sufficient statutory power would "just say no." Yet we see government biologists and hydrologists and foresters and others participating in, and their agencies supporting, virtually every watershed effort that has achieved any level of success.

When I first suggested the idea that government participation in watershed groups is vitally important, one of my friends who is close to these groups resisted and argued strongly that the effectiveness and the beauty of the watershed movement was its independence from government. Although my observations, if not my heart, led me to the contrary conclusion, I worried that in my cursory research I had not looked widely or closely enough at the groups.

Then, in late 1997, Dr. Douglas Kenney with Betsy Rieke of the Natural Resources Law Center presented a report to the Western Water Policy Review Advisory Commission (WWPRAC) concluding that: "The federal government plays a significant and essential role in the effective functioning of most watershed initiatives." They cited several cases (including some that I had looked at) where federal agencies not only provided most of the resources but were also pivotal in initiating watershed efforts. Furthermore, government has been the key to implementing the management strategies selected by the groups. Conversely, when watershed groups have foundered, Kenney attributed it more to a lack of federal support than to barriers erected by the federal government. The report recognized that many of the issues that motivate the creation of watershed initiatives are federal requirements, and therefore concluded that "the federal government is clearly part of the problem and the solution."

In its final report, the WWPRAC paid little attention to the Kenney study and, other than a nod to the impressive and promising recent record of watershed groups, did not address how the federal government could contribute to furthering the movement. Instead, WWPRAC proposed an imitation of the experiment in federally mandated river basin commissions that was tried and failed in the 1960s.

Unlike the more insightful recommendations in the WWPRAC report, this ill-considered proposal sparked widespread criticism. It was perceived as a call for big top-down commissions that would put a federal hand on all the nation's rivers. Of course, the western water establishment is chronically thin-skinned and especially suspicious of any ideas contained in a federally generated report. So perhaps attention to the WWPRAC report was destined to be negative. But the questionable proposal for river basin commissions was not only unsupported by experiences and reports submitted to the WWPRAC; it was also bound to be politically dead on arrival. Furthermore, many of the more thoughtful observations and recommendations of the WWPRAC; were diluted by including it. The WWPRAC might have enhanced the overall acceptance of its report by removing a major lightning rod from it. And the utility of the WWPRAC report could have been vastly increased by paying attention to the Kenney report.

Although experience shows a need for government involvement in watershed efforts, I fear that overformalization of the federal role could be destructive. I worry about agencies or Congress trying to concoct rather spontaneously programs and legislation to institutionalize what has worked so far. An attempt to design a federal definition or standards for watershed initiatives could undermine their effectiveness. Surely it would help to have more money in agency budgets to finance watershed efforts. But if the price of funding is compliance with detailed regulations intended to force the groups into a mold, the command for uniformity could smother local incentive and creativity and the essential flexibility of these groups.

Ham-fisted government participation can ruin a watershed initiative. For example, the Clear Creek Coordinated Resource Management and Planning Committee was dominated by the Forest Service and the Idaho Division of Water Quality. Landowners were supposed to implement land-use practices that would improve salmon streams degraded by timber harvesting and grazing and stream channelization. When too few landowners participated to satisfy the federal agencies, they nominated the watershed to be part of the state's "Stream Segments of Concern" program, a move that could lead to mandatory practices. This caused a reaction from antigovernment individuals who dominated meetings. Finally, the state agency decided to end the process altogether.

The Clear Creek watershed management initiative was driven almost solely by federal and state agencies, and so it failed for lack of local support.

Instead of trying to foster cooperative management, the agencies apparently sought out the local community to endorse their programs. By contrast, the most successful efforts that I examined were those that allowed local people to "drive the bus." A government agency may start a local initiative. But ultimately, the group must be guided by local interests, and government personnel must become equals among the participants.

The Upshot

It may be surprising to suggest that watershed initiatives, supposedly paragons of cooperation and compromise, must take on the banner of a cause. And it is odd to ask if these bastions of independent grassroots spontaneity need the partnership of governments. The implications undermine the image of groups putting collaboration above cause and eschewing the rigidity and irrelevance of governments pushing square-cornered programs.

The questions I raise do, indeed, put watershed groups in a different light. But it may not be so bad to amend our vision of collaboration to include tackling tough issues internally, making hard decisions, and ultimately reaching the point where a group can act and advocate jointly a single cause. If these groups cannot get far without government participation and support, it puts the official government participant in a more comfortable and legitimate position and helps to move the watershed movement beyond primitive antigovernment notions. And it recognizes that, at least in this realm, government itself may be disproving our old stereotypes.

The effectiveness of watershed groups will be enhanced by recognizing and taking advantage of their rich recent experience. Improving the health of watersheds is a purpose well worth the effort. Moreover, I believe watershed groups can be a moving force in fulfilling the ideal of civic virtue and participation in our society well beyond the realm of natural resources problem solving.

RESOURCES

Douglas S. Kenney et al., *The New Watershed Source Book*. Boulder, Colorado: Natural Resources Law Center, University of Colorado School of Law, 2000.

Betsy Rieke and Douglas Kenney, *Resource Management at the Watershed Level: An Assessment of the Changing Federal Role in the Emerging Era of Community-Based Watershed Management*, March 1997, available from the Natural Resources Law Center, University of Colorado School of Law, Campus Box 401, Boulder, CO 80309-0401.

Western Water Policy Review Advisory Commission, *Water in the West: The Challenge for the Next Century*, 1998, available from the Western Water Policy Review Office, P.O. Box 25007, D-5001, Denver, CO 80225-0007.

Are Community-Based Watershed Groups Really Effective? Confronting the Thorny Issue of Measuring Success

Douglas S. Kenney

Despite the efforts of numerous investigators, critical research focused on the western watershed movement still lags woefully behind the level of experimentation and promotion. In some respects, this situation reflects many of the qualities of the movement itself, including an emphasis on action and ad hoc experimentation rather than on deliberate study or standardized planning, and the fact that most western watershed groups are still relatively young, which greatly complicates their study. But if you assume, as I do, that the movement in favor of community-based resource management is the most significant and exciting development in resources management since the environmental movement of the 1960s and 1970s, then this lack of critical analysis should be at least somewhat disconcerting.

The Issue of Success: What We Don't Know

The primary question facing analysts is the most obvious: Are watershed groups effective? This seemingly simple question has many dimensions. First and foremost is the definition of success. Clearly, two different measuring scales are currently in use. The first states that success is achievement of a specific on-the-ground goal described in terms of improved environmental health. Using this definition, a group organized to restore water quality is successful only if and when water quality measurements show real improvements, such as a decline in biochemical oxygen demand (BOD) or nitrate pollution. Given that this is the primary definition utilized by critics to measure the performance of well-established governmental programs and entities, it seems fair and reasonable to apply this standard as well to "alternative" means of problem solving and management. On the other hand, many environmental problems are the result of decades of abuse or ignorance, and tangible, measurable progress cannot realistically be expected in many cases for decades. Consequently, imposing this definition of success on a community-based watershed restoration effort can be unfairly burdensome.

That argument leads to the second definition of success, which is more forgiving. This definition states that success can be measured by "organizational" parameters such as improved relations and trust among stakeholders and managers, perhaps deriving from increased communication among relevant parties. Other measures of organizational success include an expanded role for local parties in new processes of planning, management, and decision making, especially those that recognize the systemic and transboundary qualities of natural resources. Success is also presumably demonstrated by the development of processes that emphasize action over seemingly endless debate and study, and that articulate green goals such as water quality improvements, enhanced fish and wildlife habitat, resource conservation, and related goals of environmental restoration. Whether this definition is endorsed by a given party is often dependent upon whether he or she believes that this type of success is a prerequisite to achieving the more fundamental, on-the-ground form of success that results in measurable improvements in resource health.

Many parties question if this cause-and-effect relationship really exists. Building personal relationships among historic adversaries, it can be argued, encourages difficult issues to be ignored (as they might threaten newfound friendships) or, worse yet, to be addressed through inappropriate compromises. Reed Benson of Oregon Waterwatch and Michael McCloskey of the Sierra Club, for example, warn of the perils of "least common denominator" decision making that can result from consensus-based processes. They also warn that these processes, in providing mechanisms for greater local input in decision making, reduce the influence and effectiveness of the national environmental organizations in the policy arena—and where would watershed restoration be without federal hammers such as the Endangered Species Act and the Clean Water Act? Additionally, it is occasionally argued that the pro-environmental mandates of some groups is a ruse, and part of the historic concern that local decision makers will favor commodity interests over conservation objectives.

If these and related concerns—explored in detail in *Arguing About Con-*

sensus—prove to be accurate, then the second definition of success, by far the most widespread definition in current use, is largely invalidated. Some people would undoubtedly continue to cling to the definition, arguing that collaborative, bottom-up processes have normative merit regardless of their outcomes. In part, this line of reasoning is bolstered by identifying the perceived shortcomings of traditional arrangements and, in particular, their reliance on distant federal laws and regulatory programs, overly formalized decision-making processes, and adversarial decision-making processes. In contrast, the so-called watershed approach is generally viewed as operating "outside the box" of normal government, offering greater pragmatism in resource management. While many elements of that argument are indeed compelling, ultimately I suspect that most observers believe that any real definition of success must require achievement of—or real progress toward—on-the-ground goals in resource maintenance or restoration. Only if they are a means to a practical end do the "feel good" products of collaborative-based processes merit the level of enthusiasm that many parties, myself included, have expressed.

In most cases, the research community does not have sufficient evidence to support any definitive conclusions about whether this second form of success begets the first. And to the extent that success stories are identified, a further challenge is identifying which qualities, structural or functional, of a particular effort account for its success. Watershed groups, after all, are a collection of a several "institutional" reforms collectively modifying the manner in which issues are framed, decisions are made, and solutions structured and implemented. Some of these elements, such as geographic regionalism and substantive integration, are ideas with a long history of scholarly (although not political) acceptance; but others, such as the use of the collaborative/consensus model of decision making and problem solving, are elements of debatable utility in some cases. Leaving aside the fundamental political questions regarding the appropriate role of citizens and communities in public policy decision making, the collaborative/consensus model raises some real issues of practicality.

For example, I've always found it interesting that while most watershed groups express a concern both for water quality and quantity issues, the groups actively working to resolve water quality issues dramatically outnumber those actively engaged in water quantity debates. Yet, I know that water supply is not now, nor has it ever been, such a secondary issue in the region. I suspect that the explanation to this trend lies in the different set of problem-solving incentives offered by water quality restoration, which typically offers widely distributed benefits, versus water supply augmentation, which frequently only benefits the next appropriator in line. Observations such as this suggest that the collaborative/consensus component of a watershed group can be either an asset or a liability in problem solving, depending on the particular context of an effort. Similar relationships undoubtedly exist for other elements of the watershed approach, including the typical public-private composition of watershed groups and the informal and flexible structure of most efforts. Better understanding the manner in which these components interact with each other and with the natural resource problem at hand is a critical research need, especially if we are being asked to assess the prospects for success or failure of groups that are still too young to have faced a real-world test.

Why Should You Care?

The issues I discuss are of obvious interest to watershed researchers, but why should a broader audience be concerned? The answer to this question lies in a consideration of the current policy environment in natural resources, which at both the national and state levels is, in most cases, increasing pro-watershed and pro-collaboration/consensus. The work of the Clinton administration—especially through the Environmental Protection Agency and the National Performance Review—to cultivate the watershed approach is perhaps the most visible expression, but activity at the state level is also significant. This is perhaps best illustrated by a recent policy resolution of the Western Governors' Association, asserting that the "nature of environmental and natural resource problems is changing" and that "innovation solutions hold the prospect of achieving the desired environmental outcome and increasing economic wealth." Some of the key elements of these "innovative solutions," according to the governors, are replacing command-and-control programs with incentive-based systems, increasing the role of the public in problem definition and the crafting of solutions, a greater reliance on physically relevant administrative boundaries (such as watersheds), and perhaps most important, a commitment to collaborative decision making.

Personally, I think the governors are probably correct. I believe that the western watershed movement is the most exciting and significant innovation in natural resources in a quarter century. In addition, I think that the typical features of most efforts—that is, a regional focus, public-private participation, collaboration/consensus decision making, informal authority, environmentally sensitive mandate, and pragmatic focus—are right on the money. I am also convinced that the old way of doing things is often of limited value, although it is worth noting that traditional management programs, such as the point-source emissions control scheme of the Clean Water Act, have produced many significant environmental improvements. I am a proponent of the western watershed movement. But still I am troubled by the unanswered questions. For example, if this movement is good for the environment, then why are many of the "skeptics" environmentalists? Does an enhanced role for local stakeholders mean disempowering national interests? And perhaps most important, why do many of the most active researchers in this field find it difficult to identify more than a couple of good success stories, or feel compelled to attach an extensive list of disclaimers to those they do identify? I suspect we've reached a point where ignoring the fundamental questions underlying these observations does more harm than good to the interests of the watershed movement.

I very much want the watershed movement to succeed and prosper. I want to go before policy makers and tell of great success stories, and to offer effective advice about those structural and functional qualities that lead to success. I want to be able to stand before some of my more academically rigorous colleagues and say, "I have the proof you've said is lacking." And most of all, I want to enjoy the promise of community-based watershed management— namely, healthier environments, stronger communities, and less bureaucratic inefficiency. However, I can't help but wonder sometimes if the greatest impediment to achieving these goals of the watershed movement may be an "overenthusiasm" of some of its proponents. For example, establishing groups in areas lacking a critical mass of involved stakeholders, or groups focused on problems not amenable to collaborative/consensus processes, will undoubtedly increase the number of failed efforts, discrediting the entire movement. At the other extreme, seemingly successful groups that move forward too fast will undoubtedly make missteps costly to the movement. Recent controversies regarding the Quincy Library Group forest management legislation and the Oregon Plan for salmon recovery, for example, likely have done more harm than good in easing the concerns of critics wary of increasingly common proposals to bring greater formal authority and structure to community-based watershed efforts.

Conclusions

I have resigned myself to the fact that I am too much of a skeptic to satisfy the ardent proponents, and too much of a proponent to appease the skeptics. While I like to consider my stance one of guarded optimism, many will see me as just a fence sitter, straining for additional opportunities to jump into the yard of the watershed disciples, but unable to totally break free from the pull of those academics demanding tangible measures of success, from those environmentalists wisely cautious of abandoning processes of environmental decision making that have produced many notable gains, and to a lesser extent, the claims of some local governments fearful that an important level of government is again being unduly bypassed.

Through this essay, I hope to convince additional proponents of the watershed movement to join me in acknowledging a need for more critical analyses and, in the meantime while such work is conducted, in advocating a more tempered set of policy recommendations. For example, we already have sufficient knowledge to stress the need for innovative arrangements in resources management and, more generally, to articulate the value of cautious experimentation and learning-from-doing. In addition, I think we can continue to encourage policy makers to steer funding to the most promising efforts. But there is no compelling justification at this time for abandoning management tools that, in certain circumstances, continue to serve us well. Similarly, efforts to formally transfer decision-making responsibility and authority to collaborative groups are, at best, premature. True, the current incarnation of watershed-based resource management is a potentially powerful new tool for resources management and problem solving; however, the true extent of that potential has not been critically explored. Until that occurs, the watershed movement is unnecessarily making itself vulnerable to threats, both internal and external, deriving from a lack of knowledge and honest self-appraisal. A more balanced mixture of advocacy and analysis is the next step forward.

REFERENCES

Reed Benson, "A Watershed Issue: The Role of Streamflow Protection in Northwest River Basin Management," *Environmental Law* 26:1, 1996.

Douglas S. Kenney, *Arguing About Consensus.* Boulder: Natural Resources Law Center, University of Colorado School of Law, 2000.

Douglas S. Kenney et al., *The New Watershed Source Book.* Boulder, Colorado: Natural Resources Law Center, University of Colorado School of Law, 2000.

Mike McCloskey, "The Skeptic: Collaboration Has Its Limits," *High Country News,* May 13, 1996.

Ownership, Accountability, and Collaboration

Andy Stahl

My wife and I farm 44 acres in rural Oregon. Bordering our farm on two sides are managed Douglas-fir forests owned by Giustina Resources, a family-owned timber company. Our house sits immediately adjacent to the property line, on the other side of which Giustina plans to selectively log. However, to conduct its logging operations efficiently, Giustina must cross our property. The stage is set for a collaborative partnership. We agreed with Giustina to do a lot-line adjustment that gives it access to an old skid road that passes through our property while, in return, we obtained a 100-foot buffer around our homesite. My wife and I got what we wanted: the peace of mind of knowing that the beauty of our home would be preserved. Giustina got what it wanted: economical access to manage 30 acres of commercial timber. The entire negotiation, from the presentation of the problem, to the formulation of alternatives, analysis of those alternatives, dickering over the details, culminating in the selection of a mutually agreeable plan, took an hour.

What made our collaboration so successful? Ownership. My wife and I chose to relinquish an acre of our property, in which no one else had a vested interest. Giustina chose to give up an asset, a like amount of its property, in which no one else had an interest. Our agreement would not cost anyone but the parties to it. On the other side of the ledger, the benefits of our agreement were spread more widely. Giustina's new access is environmentally preferable to its other options, which would have required a new skid road be constructed, probably within a riparian area to the detriment of local wildlife. The cost-benefit ledger of this small transaction illustrates a key to successful collaborations. The costs must be borne voluntarily by those sitting at the table, although the benefits can be more widely spread. Obviously, if Giustina had to cross two property boundaries to reach its land, we would not have reached agreement without the third property owner at the table.

With ownership come boundaries. Whether it is intellectual property ownership in software or real property ownership, boundaries are fundamental. The property boundaries of Microsoft Word are its particular code, including any patentable concepts. Use of that code or concepts without Microsoft's permission violates those boundaries. The boundaries of real property are its metes and bounds, the crossing of which without permission from the owner is an act of trespass. I can no more grant permission to you to enter onto my neighbor's land than she can allow entry onto mine. With ownership also comes tenure. Tenure creates incentives to develop relationships across shared boundaries. Giustina and we knew that a successful resolution of the access issue would pave the way for more successful negotiations in the future, and the reverse is true, too.

When collaboration enters the realm of public land and resources, where ownership is communal, strange things happen. No longer do the land's owners have the authority to cut deals; instead, bureaucrats hired by land management agencies control how land-use decisions are made. Though bureaucrats have substantial authority, it is virtually impossible to hold them accountable for bad decisions. What bureaucrats lack in ownership of the property, they make up for in their sense of ownership in the bureaucracy. Two examples illustrate well the difficulties authority without ownership or accountability creates for successful collaboration. The Quincy Library Group (QLG) is well known for its effort to bring together diverse local community interests, and its story is told in greater detail on page 79. What isn't as well publicized is the reason QLG sought federal legislation to implement its vision. At its inception, QLG leaders assumed that if they were able to reach consensus, the Forest Service would be pleased to follow a plan agreed to by the formerly warring factions. Nothing could have been further from the truth. In fact, the Forest Service sought to torpedo the fragile compromise by seeking to sell timber from lands the QLG participants had agreed should be off-limits to logging. QLG's timber industry supporters, in a show of solidarity, refused to bid on the errant sales, which the Forest Service withdrew only after considerable congressional pressure. The Forest Service's hostile behavior forced QLG to take its case to Congress, which was delighted to reward the local collaboration.

Why did the Forest Service behave so irresponsibly? The Forest Service isn't talking, but one might speculate that its interest in profiting from salvage sales, whose receipts the agency gets to keep and spend without congressional oversight, might provide sufficient justification to sell trees notwithstanding QLG's best-laid plans. Higher-ups in the agency hold its land managers responsible for meeting salvage sale targets because about 30 percent of the sales' receipts are spent on mid- and upper-level bureaucratic overhead. Fewer salvage sales mean fewer dollars and threats to civil servant job security.

A second example illustrates what can happen when an agency bureaucrat bucks internal pressures and embraces collaboration. Ashland, Oregon, obtains its municipal water from forests managed by the Forest Service. Decades of fire suppression have made the watershed prone to catastrophic burning, which would increase stream sedimentation. So, too, would excessive logging and road-building on sandy soils prone to landslides. The Forest Service's original plan to create fuel breaks by strategic clear-cutting received little community support. Enter Linda Duffy, the new Ashland District Ranger. Duffy encouraged a collaborative group of citizen interests that reworked the Forest Service's plan into one that focused on brush removal, thinning of small trees, and the occasional removal of a large, old-growth tree where infections of dwarf mistletoe and other diseases created hazardous conditions. Ashland's mayor commented on the success of the collaboration: "She's worked well with citizen groups and they've given her high praise."

Her forest supervisor, Mike Lunn, thought otherwise. In a surprise move forty days before his retirement from the Forest Service, Lunn summarily removed Duffy from her district ranger position the day before the collaborative group's plan was released for public review. After the group handed Duffy its plan for Ashland's watershed, Duffy stunned its members: "I've been removed from my job. I learned about that yesterday," she said. The collaborative group's mediator replied, "We're certainly not pleased. We'll be discussing how we can respond to this action. It's very disruptive to our process. . . . You did more than just talk. You put yourself on the line. We'll go to bat for you in any way we can." A high-ranking Forest Service official explained Duffy's removal, "She didn't fit her boss's mold of a Forest Service leader. She was too committed to working with people rather than protecting the Forest Service's power base."

Will Mike Lunn or the QLG's Forest Service tormentors ever be held accountable for their intransigent opposition to citizen collaboration? Probably not. In the last analysis, bureaucracies protect their own, especially when their own are protecting the bureaucracy's vested interest in retaining its authority. Regardless of the well-intentioned efforts of the Linda Duffys, agencies are loath to share power with the unwashed masses.

Politicians, however, are thrilled with collaborative processes. Successful collaboration over public resource allocation means that politicians don't have to divvy up the baby, a slicing that rarely satisfies anyone (monarchs are not subject to plebiscite and thus can politically afford to offend by splitting babies). Because forest issues are so potentially explosive politically, Congress is loath to deal with them in any straightforward manner. Advisory boards have long been a favored mechanism of legislatures and government executives to lessen the heat. In 1906, Forest Service founder Gifford Pinchot created grazing advisory boards to help quell the range wars that pitted cattlemen against shepherds, and both against Pinchot's corps of forest rangers. Pinchot was pleased with the results, noting "[a] marked improvement in sentiment among stockmen." Congress rubber-stamped Pinchot's grazing advisory boards in the 1950 Granger-Thye Act, while refusing stockmen's entreaties to reverse the Forest Service's policy of limiting livestock grazing damage.

Federal land management agencies have appointed scores of citizen councils since Pinchot started the ball rolling. They share two common attributes. They have no real authority, and, notwithstanding their impotence, they are distrusted by anyone who doesn't hold a seat on them. The outsider's antipathy toward citizen councils is well summarized by the Southern Utah Wilderness Association: "The Utah BLM should abandon its practice of privatizing public lands planning through private advisory groups. We suspect BLM favors private advisory groups as they allow the agency to avoid responsibility for difficult decisions. If BLM managers can't stand the heat, they should quit, rather than collect a paycheck while leaving the decision making to private interests."

But what if citizen councils had real authority over public land decision making? Would that spell the end of a 100-year experiment in federal authority over vast expanses of public land? Would it create the conditions of ownership, boundaries, and tenure necessary for truly effective collaboration? And would it be constitutional? In the coming half dozen years, we'll probably find out. In late 1999, in the most significant public land law initiative since the 1907 creation of the Forest Service, the U.S. House of Representatives passed HR 2389, the Secure Rural Schools and Community Self-Determination Act, which would create citizen collaborative groups for every national forest. The fifteen-member local committees would include at least one "local resource user," "environmental interest," "forest worker," "organized labor representative," "elected county official," and "school official or teacher," drawn exclusively from the geographic region of the national forest.

Unlike their federal advisory predecessors, HR 2389's committees would exercise substantial authority over national forest budgets by controlling how 76 million federal dollars annually are to be spent on national forest projects.

Over time the committees could have veto authority over an additional $500 million of the Forest Service's budget. Together with local counties, the citizen committees would make grants to the Forest Service to finance agency expenses associated with carrying out projects approved by the county and committee. Given the considerable authority enjoyed by HR 2389's local committees, it is worthwhile asking to whom they would be accountable. Committee members would be selected by the secretary of agriculture. Members would serve two-year terms, subject to unlimited reappointments. Committee members who accommodate the Forest Service's spending priorities will likely be rewarded with reappointments; those who do not can expect to see their tenures cut short. Thus these committees would not be selected through any representational process, yet they will enjoy budgetary authority previously exercised exclusively by elected members of Congress, raising issues of unconstitutional delegation of legislative function.

Regardless of whether HR 2389 becomes law, its House passage with little debate reaffirms Congress's interest in seeing public resource issues resolved elsewhere than in Congress. When local citizen groups replace democracy, owners of public resources—that is, all of us—have less say in how those resources are managed through our vote, our letters to elected officials, and the other avenues we use to influence the political process. Politicians get to pass the buck: "If you have a problem with how your local national forest is managed, take it up with the duly appointed citizen committee."

It is precisely the "duly appointed" nature of HR 2389's committees that threatens democracy and distinguishes these committees from the Quincy Library Group and other ad hoc assemblages. QLG participants used collaboration as a political tactic to build support for their agenda. That the tactic was successful is no reason to condemn QLG, per se. In fact, QLG's political success reiterates a fundamental lesson for all activists—the broader one's coalition, the more likely one's success. So long as government-appointed citizen committees had no authority, their threat to democracy was limited to the outsider's view that committee participants enjoyed undue access to decision makers (a charge that most advisory committee members would refute in frustration by citing their lack of influence!). But when government creates collaborations vested with authority and power previously held by elected representatives, as does HR 2389, that is when collaborations threaten democracy.

Communal ownership is messy, especially at the federal level where ownership is claimed equally by those far removed and those close to the land. The Constitution vests sole authority for public land decision making in the hands of elected officials in Congress, which has given over the day-to-day administration of these lands to professional land managers, subject to scores of policy directives and legal limitations. If the owners don't like the results, the voting booth is where accountability is exercised. So long as collaborative processes lack accountability to all of the public land's owners, collaborations should be entrusted with no more than advisory functions.

REFERENCE

David Schoenbrod, *Power Without Responsibility,* New Haven: Yale University Press, 1995.

Exploring Paradox in
Environmental Collaborations

Jonathan I. Lange

Compromise has been termed the lifeblood of American politics, a central identifying characteristic of democracy, even a central feature of the American character. It is also associated with selling out, refusing to take a stand, settling for less. Paradoxically, compromise is a term that can be used to commend or condemn. The same is true for collaboration. Collaboration is held as an admirable conflict resolution goal, a state of affairs toward which we should strive, as well as a signal that political actors have lost their integrity, as "collaborators" did in World War II. These dual meanings reveal the inherently paradoxical nature of alternative dispute resolution (ADR) processes.

Scholars in the field of communication define paradoxes in group life as those situations that members experience (consciously or unconsciously) as both contradictory and circular, noting that the situation allows no escape from the resulting dilemma. For example, the paradox of "individuality" is expressed by the statement that the only way for a group to become a group is for its members to express their individuality, and that the only way for individuals to become fully individuated is for them to accept and develop more fully their connections to the group. Such paradoxes can paralyze a group, preventing it from moving forward in solving problems and achieving its task. Individuals come to groups looking for what they can get; yet upon arrival, they so often receive the message: "You can't get anything here until you give something!" Further, it seems you have to join before you can tell whether joining is even a good idea.

Paradox, I argue, defines the environmental partnership experience, and only by understanding paradox can we fully appreciate the complexities of environmental collaborations. Here I explore the implications of five paradoxes in the context of the Applegate Partnership in southern Oregon, where I served as a cofacilitator for more than two years. (The Applegate story is told in greater detail on page 102.)

The Entry Paradox

The first paradox is perhaps the most fundamental, since disputing parties encounter it as soon as they enter the dispute resolution arena. Although ADR processes emphasize voluntariness as a defining characteristic (as opposed to forced participation, such as might happen in court), it is surprisingly difficult to find instances of unpressured use of the process. Few disputants enter mediation by spontaneous mutual choice. Instead, most instances involve reluctant parties entering the process under strong social pressure, under pressure mobilized by the other disputant, or because they are required to do so by a government agency. Environmental collaborations share this entry paradox. For example, one of the original environmental representatives in the Applegate Partnership told me he was there mainly "to keep an eye on things . . . originally, we were there to see what was going on, and to make sure nothing happened [that would hurt our cause]." Timber industry representatives also felt pressured to enter. One member said he decided to participate because he felt there was "a gun to my head and to the head of the whole industry." How will people act in a group when they are pressured to be there?

Psychological Reactance

It is difficult to infer direct behavioral consequences resulting from a paradox, especially if the paradox is experienced unconsciously. Nonetheless, I believe that the theory of psychological reactance is a useful frame for explaining some of the consequences of the entry paradox, and perhaps other paradoxes as well. J. Brehm defines psychological reactance as a "motivational state" consisting of internal pressure directed toward reestablishing a freedom perceived to have been lost. A derivative of cognitive dissonance theory, reactance theory posits that a person becomes "motivationally aroused" any time he or she thinks a personal freedom has been threatened or eliminated and that in such cases, the person is moved to attempt to restore the freedom. The theory helps explain widely experienced phenomena such as "the grass is always greener," and "buyer's remorse," or any situation in which a person perceives an actual or threatened loss of freedom to choose. While both the magnitude of arousal and the effects on an individual's behavior vary, attempts to recover a freedom tend to persist until successful or unless reactance is reduced in some other way.

Forced entry into ostensibly "collaborative" processes might elicit such reactions. If representatives in an environmental collaboration experience the process as something in which they *must* participate, the theory would predict that they would act in ways designed to recover their freedom. Brehm further notes that the behavioral effects of psychological reactance include either attacking a person or persons perceived as responsible for the loss of freedom or defiantly exercising the freedom itself. This could explain some of the many personal attacks and threats to quit the process that characterized a number of Applegate meetings. It's not difficult to imagine "acting out" when one feels forced to be present, especially with people with whom one would rather not share time.

The Authority Expropriation Paradox

The second paradox, which I call authority expropriation, is a paradox of power, and calls into question the entire existence of environmental collaborations such as Applegate. In collaborations involving public lands, partnerships are established to address knotty land management questions, but the actual power to make decisions remains with agencies such as the U.S. Forest Service and the Bureau of Land Management. This paradox produces tensions among agency personnel and partnership participants alike. For example, Applegate members often complained that their supposed "empowerment" by the agencies was nothing more than symbolic. Frustrated at their lack of decision-making authority, they frequently railed against agency inaction or unwillingness to enact their suggestions and recommendations. Meetings were characterized by regular calls for "getting the agencies off their asses." While partnership members felt misled, let down, and even "disempowered," at least some agency members began to resent the partnership for attempting to take too much control. After all, the agencies had final decision-making jurisdiction and these groups were trying to expropriate their lawfully assigned role and legitimate authority. Some personnel within the agencies complained about the extraordinary time required to respond to issues raised by partnership members, and began to resist partnership initiatives. After attending a number of partnership meetings, another agency member said "the lack of expertise" there was appalling; he began to hold in doubt any information the partnership generated. A different agency member said that some colleagues resented the partnership, seeing it as "just another interest group." A third member worried that "We're already seen as puppets of the partnership." Finally, agency personnel would get particularly annoyed when partnership members would take field trips, and upon finding some element of land management wanting, accuse agency personnel of not doing their job properly, or worse, forcing partnership members to do it themselves. Such frustration and resistance are clearly connected to the paradoxical situation created when lines of authority blur.

The Stakeholder Paradox

A third paradox intrinsic to the vast majority of environmental collaborations concerns the critical issue of who gets to be a stakeholder. Barbara Gray defines stakeholders as those parties with an interest in the problem, including all individuals, groups, or organizations that are directly influenced by actions others take to solve the problem. Nearly all ADR theorists and practitioners agree that it is vital that all stakeholders in the process be present or at least represented in order for partnerships to succeed. Partnerships must bring relevant expertise to the table, develop "ownership" of the problem, generate the ability to implement emerging solutions, and avoid the possibility that excluded stakeholders might later sabotage the process or solution.

But identifying stakeholders in environmental collaborations can be especially problematic and paradoxical. The stakeholders who represented the "environmentalist" contingents in the Applegate Partnership were thoughtfully and carefully chosen. But because many partnerships involve national lands, it could be argued that a representative such as Chris Bratt—a twenty-year Applegate Valley resident, and longtime board member of the major environmental organization in southern Oregon, Headwaters—is representing more than just his local environmental constituents. In a sense, Bratt also represents the members of national environmental organizations (i.e., the Sierra Club, the Wilderness Society, the National Wildlife Federation, and others), since they too have a stake in the management of the national forests within the Applegate watershed. Pushing this perspective further requires us to consider at least the idea that since this is national land, "owned" by the citizenry, Bratt represents millions of stakeholders who may not belong to an environmental organization, but who might care or be affected by how the land is managed. Going further, one might even suggest that this paradox holds true in air quality disputes, river use conflicts, decisions about energy use and toxic materials disposal, and just about any environmental controversy that has significance beyond the particular place in which it is debated. Some environmentalists have even argued that environmental constituencies include wildlife, trees, and other flora and fauna.

While this begs the entire question of stakeholder representation, clearly it is a paradoxical issue, as conveners and participants must make choices about representation that cannot possibly satisfy the full stakeholder participation requirement. In the Applegate Partnership, the practical effect of this paradox was that almost anyone could come to the table and claim the status of stakeholder. And this they did, as the partnership's declared intent to remain inclusive and community-oriented evoked visits from not-so-cooperative interested parties—stakeholders both inside and outside the community—who tended to impede the process. These stakeholders often put forth unrealistic ideas, proposing, for example, an innovative but unusual form of air-crane logging that took several weeks for the group to consider. They also required the group to go over old ground as the newcomers were "brought up to speed," often a time-consuming process. They disrupted the general cohesiveness of the group as they sometimes engaged in behaviors that emphasized differences and conflict. One example was the tendency to remind group members of past injustices committed by the "other" party. The stakeholder paradox makes already difficult decision making even more demanding, if not impossible, as it undermines forward movement by the group. This particular quandary is related to the fourth paradox that plagues environmental collaborations.

The Constituency Paradox

The constituency paradox concerns a bizarre set of inconsistencies that plague environmental collaborations. While all ADR collaborative efforts seek to find common ground between conflicting interests, environmental conflicts do not cease, temporarily or otherwise, during the ADR process. In environmental conflict, parties' *competitive* behavior continues simultaneously—right through the *collaboration* that is being attempted at the negotiating table. For example, as representatives from the timber industry, environmental groups, and resource agencies were attempting to collaborate on management of the Applegate watershed, constituency members from both the industry and the environmental camps concurrently were filing lawsuits against the agencies for allegedly illegal forest management practices in the Northwest. In addition, both environmental and timber activists continued intense negative information campaigns against each other. They lobbied the public, Congress, and agencies in an essentially competitive strategy, while on the Applegate—as well as in other partnerships in the West—timber, environmental, and agency "representatives" were attempting to find collaborative solutions to localized forest management problems.

While such a situation is not uncommon in larger political affairs—as when trade sanctions are enacted between allies—this paradox haunts environmental collaborations. The Applegate Partnership's progress was regularly slowed each time news of some constituent's recent tactic would reach representatives. Tempers frequently flared as collaborating partners were angered at each other's constituent behavior, whether it was lawsuit initiation, unflattering portrayals in the press, lobbying of public officials, or other aggressive, competitive strategies.

The constituency paradox has significant implications for what might be called "intraconstituency communication." Communication between different constituent members revealed deep differences about the inherent value of environmental collaborations in general, as well as the Applegate Partnership in particular. Arguments about even participating in collaborations went beyond the usual ambivalence or intraconstituency disagreement in labor-management negotiations. Here, within each constituency—whether that be timber, environmental, agency, or community member—there were unending arguments about foundational issues such as whether or not the partnership was worth representative and constituency time, whether or not anything useful would ever come of it all, the legality of the partnership, the likely quality of any decision the group might make, and so on. Since I spent more time with the environmental constituency than with others, I offer the following illustrations of the arguments environmental constituents engaged in, though similar kinds of feuds no doubt plagued other factions. The illustrations are offered to indicate how deeply the paradoxes of environmental collaborations can reach.

Negotiating the Constituency Paradox on the Applegate

Within the space of a year, the Applegate Partnership became the anvil on which the environmental community's internal politics were hammered out. Headwaters, the local forest-oriented environmental group, originally went out on a limb for the partnership, putting the environmental group in a new role as a relatively moderate forest activist. Local, on-the-ground experience compelled Headwaters to advocate for the thinning of small trees in order to improve forest health and prevent catastrophic fire. The group adopted a "forest management" or "conservationist" stance. This position contradicted more radical environmentalists who argued that all logging was to be fought with whatever means available. At the time, these so-called radicals held a "zero-cut" position, more akin to a "preservationist" orientation. When I asked Headwaters President Julie Norman if her organization experienced pressure from the more radical groups, or from the nationals concerned about local control, at first she demurred. Later she admitted: "We've been bucking the tide on this from day one. And we defended it; we thought we could do it; we hoped it would work out, you know, we really did and we were just prepared to take the flak." Nevertheless, Headwaters eventually withdrew its formal participation in the partnership.

While the press and other environmental groups criticized Headwaters for leaving, Headwaters staffers, to a person, explained their defection as the result of internal communication and relationship issues, not external pressures. The group's president cited time requirements, lack of clarification of partnership processes, and the fact that

> We didn't have our own shit together. There were these areas of fuzziness in our policies. We weren't sure just how much leeway to give Chris or Jack. And when Jack was going all over the country saying all these things that we couldn't agree with . . . like "When people come to the Partnership meetings, they leave their constituencies behind," that kind of thing, it just drove us nuts.

At first glance, all this might seem part and parcel of any constituent group's internal politics, most notably the predictable desire to keep representatives on a short leash. Other Headwaters staffers, however, were more dramatic in their explanations of the organization's desertion. The intensity of their accounts seemed to stem from the special difficulty created when they found themselves with representatives working toward collaboration while they were spending their time attempting to thwart the industry and/or resource agencies—the constituency paradox. These staffers told me that the partnership was dominating their agenda, the philosophical disagreements were incessantly and unproductively argued, anger was commonly expressed; decade-long friendships and working relationships were nearly abandoned. One alternate partnership board member and environmental representative from Headwaters claimed that Headwaters had been "poisoned" as a result of the partnership. Another Headwaters member said the disagreements "almost broke us up."

Over time, it appeared that environmentalists were fighting with each other almost as much and as ardently as they did with the industry. One might hypothesize that the more the disputants collaborated, the more their constituents would internally conflict. At the first West Coast Ancient Forest conference, held in February 1993 (shortly after the Applegate Partnership formed), I heard speeches, panel presentations, and discussions about "involving the community" and bringing organized labor into the environmentalist camp. There was a spirit of hope as science and the courts seemed to be lining up on environmentalists' side. It was at this first conference where I first saw the prototype artifact of the then-embryonic Applegate Partnership: a button with the word THEY circled and lined out in red; that is, the (internationally signed) negation of any "they." However, one year later, at the second conference, some activists wore competing buttons; the word TRUST was negated with a red line.

The second conference was dominated by disagreement between those who favored preservation, zero cut, and nationalization of the issue and those who argued for conservation, thinning, and local community-building. Activists repeatedly asked themselves, "Can we be watchdogs and participants at the same time?" Several activists who had participated in collaborative attempts that had "failed" spoke vehemently against partnerships, warning of cooptation, compromise, and a "coercive harmony." "It's our public lands," argued Tim Hermach, nicknamed the Godfather of Zero Cut. "We wouldn't take the treasure out of Fort Knox or the Smithsonian to create temporary jobs. Why should we do it in the Pacific Northwest?" Others who had positive experiences with other partnerships talked of "forest health" and "transforming communities." They warned of potential losses and political backlash that radical positioning would cause. Eventually, these recurrent arguments surfaced in the press with headlines such as "Disagreement on Consensus: Applegate Model Splits the Greens."

By early 1995, at the third annual West Coast Ancient Forest Conference, the more radical voices argued for "retaining the moral high ground" (by refusing to compromise); they insisted that zero cut was politically defensible. Drawing a continuum of political positioning on butcher paper, one speaker contended that "Zero cut should be seen as the middle position by the [Clinton] administration. No cut on private land should be our next move. The further to the left we go, the further to the left we pull the middle." This strategy, recalling the radical positioning of the environmentalist group Earth First!, was obviously at odds with those who defended partnership efficacy.

In the end, the constituency paradox evokes an agony that environmentalists must face in elevating either ecological values—accompanied by a relatively "pure" environmentalism—or civic values, reflected in an environmentalism marked by compromise, community, and democratic practice. The difficulty of the choice raises the question of environmental activists ever participating in partnerships. Such musings call into question the entire issue of representation and how it is accomplished, and thus the nature of collaborations in general. This kind of questioning leads to consideration of the fifth and final paradox.

The Mainstreamer Paradox

The mainstreamer paradox is related to the foregoing in that its roots reside within the representative-constituent relationship. It differs from the above in that this paradox is experienced by representatives at the bargaining table, not by constituents who are absent. In a way, it is the antithesis of what is sometimes described as "the radical's paradox," which unfolds when a radical activist thoroughly rejects a set of political practices and articulates a substantially different agenda and set of alternatives. Radical discourse can follow only one of two paths, both unsatisfactory. The discourse is either dismissed as impossible to achieve or it is fully considered, in which case radical arguments are inevitably subsumed within the larger context, reinforcing the current set of practices. Espousing radical change in any context thus places an activist in the paradoxical position of either facing dismissal or legitimating the context that he or she would otherwise destroy.

The mainstreamer paradox is like the radical's paradox, except in reverse. No matter what collaborative participants and conveners do to include all those who have a stake in the outcome (recall the stakeholder paradox), radical activists will either refuse a seat at the table, labeling the process as bogus and coopted, or they will join the partnership, undermining and sabotaging it as they participate. By definition, radicals must boycott or subvert any mainstream process, especially one that attempts to include them by deeming them representative or constituent stakeholders. The paradox then is that for environmental collaborations to achieve full stakeholder participation, a factor often seen as required for collaborative "success," they must include, even if indirectly, those who will attempt to undermine that success.

Of course, radicals are rarely invited to the table, which simply reinforces the contradiction between collaboration theory and practice. Even though collaborations must include all interested stakeholders, disputants chosen as representatives usually go through some kind of screening process, usually conducted by mediators or conveners. This is done to improve the likelihood that agreements can be reached. While this seems reasonable, given the possibility of unskilled or ill-prepared negotiators (let alone negotiation "saboteurs"), even self-selecting assessments tend to eliminate intemperate voices from the table. These voices often come back to haunt collaborations as they move from process to implementation.

In the end, the Applegate Partnership did not bring peace to the forests, as initial hyperbole in the regional and national media promised. Instead, Applegate produced more paradox than peace, but there is much we can learn from this experience. Wherever they may be located, environmental collaborations must deal with the consequences of paradox and its sometimes vicious circularity. As in other contexts marked by paradox, participants may experience a number of debilitating subjective effects. Felt loss of freedom, usurpation of choice, confusion, anger, and frustration all haunt different partnership members at different times. The behavioral consequences range from member resistance, verbal attack, withdrawal, and even damaging relations with previously sympathetic constituents. Perhaps the best way for groups to address the problems resulting from paradox is first to acknowledge, as opposed to avoiding or attempting to eliminate, the inevitable paradoxes that characterize environmental partnerships.

REFERENCES

Bingham, G., *Resolving Environmental Disputes: A Decade of Experience*. Washington, D.C.: Conservation Foundation, 1988.

Brehm, J., "Responses to Loss of Freedom: A Theory of Psychological Reactance," in J. W. Thibaut, J. T. Spence, and R. C. Carson, eds., *Contemporary Topics in Social Psychology*. Norristown, N.J.: General Learning Press, 1976.

Friedman, R. A., "The Culture of Mediation: Private Understandings in the Context of Public Conflict," in D. Kolb and J. M. Bartunek, eds., *Hidden Conflict in Organizations: Uncovering Behind the Scenes Disputes*. Newbury Park, Calif.: Sage, 1992.

Gray, B., *Collaborating: Finding Common Ground for Multiparty Problems*. San Francisco: Jossey-Bass, 1989.

BROADENING
ENVIRONMENTAL HORIZONS

Perhaps the greatest promise of collaboratives is that they open new avenues for citizen participation in natural resource management decisions. But these avenues are not necessarily open to everyone automatically. Achieving full participation remains an elusive goal for the collaborative conservation movement, as some sectors of society remain outside the movement or find it difficult to participate effectively. Essays in this section explore the challenges associated with "hearing all the voices" in collaborative settings.

We begin with Luther Propst and Susan Culp's proposal for new environmental strategies that appeal to humanity's imagination and desire to live well in the world, and approaches that tap into the power of place. A key principle supporting this strategy, he argues, must be compassion for both nature and diverse people and communities. But as essays by Beverly Brown and Maria Varela demonstrate, this is easier said than done when collaborative groups enter terrain long divided by bitter historical injustices, language and class barriers, and divergent cultural expectations. Although Beverly Brown's story of forest workers in the Northwest offers more hope for moving across the "great divide" than Maria Varela's account of cultural confrontation and misunderstanding in New Mexico, it is important to keep in mind that collaborative processes offer no panacea for conflicts that reach deeply into our cultural identities and experiences.

But if any political venue is capable of helping us work through our differences constructively, it is difficult to imagine a more effective process than local, face-to-face conversations and problem solving. All the voices must be present, and they must be weighted in proportion to the contribution they can make toward a shared, sustainable future. Again, this is easier said than done when some voices seem to have a privileged position in the conversation, and other voices don't seem to be at the table at all. Essays by Jim Burchfield and Shirley Solomon address these issues. Concerned about the hegemonic and often misunderstood voice of science in natural resource management processes, Jim Burchfield argues that science should be just one of many voices in these processes. Similarly, Shirley Solomon, in a moving essay that summarizes the essence of this book's message, encourages us to hear the voices that are not directly at the table—the voices of salmon, our own intuition, and our innermost longing to do better by a world of countless beautiful places.

As all the essays in this volume demonstrate, collaborative conservation is still a work in progress. No process or movement can solve all our problems. We hope that the stories and analysis we offer here point to an important "third direction" in environmental strategy, even though we recognize that there is not necessarily a single "third way." Rather, there is a demonstrable collection of ways based on local histories, conditions, and constituencies. As a group, these new approaches are amenable to ideas and processes that signal a marked change from environmental advocacy of the past three decades. And, while we do not see collaborative conservation as a replacement for traditional environmental conservation approaches, we do believe that collaborative initiatives will enrich, deepen, and make more effective the environmental movement as a whole.

Imagining the Best Instead of Preventing the Worst: Toward a New Solidarity in Conservation Strategy

Luther Propst and Susan Culp

As Donald Snow observes in the introduction to this book, the conservation movement has had considerable success over the past thirty years in broadening public recognition of environmental concerns and the importance of protecting natural resources and wild areas. This public acceptance has led to successes in establishing basic national standards to protect air and water quality, to control toxic substances, and to clean up polluted sites. It has also led to the designation of more than five million acres of the nation's land as wilderness; the expansion and improved management of our national parks, wildlife refuges, and conservation areas; and protection for threatened and endangered plants and animals.

While much has been accomplished, much remains to be done. Time will be the judge of how effective existing environmental protection will be in the face of the continued juggernaut of population growth, consumerism, urban sprawl, and the ever-widening footprint of extractive industries. It is clear, however, that the environmental movement, with all its diverse voices and approaches, has earned a seat at the table as a political force. Today, conservation interests can exert, in most but not all venues, an effective veto over the most environmentally egregious ideas. As shown in recent Congressional sessions, environmental advocates seem capable of rallying the public to counter efforts to roll back the current environmental safety net.

With this success, however, has come a growing estrangement from many sectors of the public who logically should be conservation's allies. These sectors include rural communities, where many residents support conservation goals, but have trouble with the usual methods of realizing these goals, which seem to favor suburban values. This disconnect also exists within the growing number of green businesspeople who support conservation, but are completely turned off by overly bureaucratic and regulatory approaches. American Indian nations and communities of color, as well as inner-city residents profoundly affected by and concerned with environmental quality, are also often alienated by the mainstream of the environmental movement. Also, a significant and growing number of ranchers and landowners, who genuinely support conservation goals, are increasingly estranged from the mainstream environmental movement.

To address this challenging situation, environmental advocates in North America must better incorporate the following three principles if we are to solidify our current victories and increase our effectiveness in coming decades.

1. Articulate a compelling vision for the future that appeals to the hopes and aspirations of the public. The environmental movement must articulate a compelling vision for the future with an appeal that transcends its traditional, but relatively small, base of suburban middle-class and upper-middle-class supporters. Carrying out this vision will require moving beyond the current emphasis on primarily defensive actions—enforcing national pollution standards, setting aside and defending what are still generally small and isolated wilderness areas, and stopping the most destructive and high-profile development proposals. It will require putting more resources into reducing the polarization and disenfranchisement that threaten the long-term sustainability of our conservation victories. Unless we figure out ways to better integrate conservation into the lives and well-being of a broader cross section of the public, our conservation victories will indeed be short-lived.

Commentators and critics of the environmental movement from across the political spectrum note that the moralizing tone of many environmental advocates has soured in its popular appeal. From the left, Mark Dowie, former publisher and editor of *Mother Jones* magazine, for example, quotes writer Stephanie Mills in describing "deep ecology" as a "puritanical approach, massively unappealing, [which] guarantees only a tiny following of visionaries and masochists, folks who function best when guilty." Dowie also describes the splintering of the environmental movement as conservationists continually try to "out-purify" one another in the popular media. According to Dowie, "a dozen separate and conflicting environmental ideologies cannot form or inspire a movement. American environmentalism needs a synthesis of disparate and apparently antithetical ideas—an indigenously American ideology that can be comfortably embraced by major institutions, academic disciplines, economic philosophies, and religions."

The media, of course, prefers to emphasize conflict, polarizing rhetoric, and oversimplification of issues into convenient sound bites. This leads many environmental advocates to conclude that the best way to bring attention to themselves or their cause is to offer the most adversarial and even outrageous quotes possible. We then mistakenly equate being quoted in the popular press with a meaningful conservation victory. The overall result is that the public increasingly equates environmentalism with "anti-ism," a movement defined by what it is against, not what it favors.

The fundamental goal of the environmental movement must be to transform the hearts and minds of the public and of society, not simply to win battles in Congress or the courts. The desire of Americans to receive positive messages can be found in their rejection of political mudslinging. To move beyond the plateau that the conservation movement has reached, we must identify effective means to transform social and personal goals in order to reflect a deeply ingrained set of conservation values. The consistent message from the environmental movement should be positive, healing, and unifying. It should inspire the public and appeal to humanity's highest ideals, to our imagination for a better world, and to our sense of justice—not only among generations but among diverse peoples as well. The task that remains for the conservation movement is to articulate a vision for society that shines brightly enough that people are inspired by its cause.

2. Demonstrate compassion for both nature and people, not only for nature. The environmental movement is subject to increased scrutiny and criticism from the public for a perceived lack of understanding of and compassion for diverse peoples and communities. This results from several factors, some legitimate and some not. We must differentiate the rhetoric of people who do not have the best interest of the environment and wild areas at heart from the legitimate skepticism of well-meaning people who have concluded that the environmental movement lacks compassion for rural people, people of color, people from poor Central American countries who migrate north to find a better life, and people who make their living—for better or for worse— from economic activities that are not sustainable and not environmentally benign. We must not expose ourselves to the allegation that the environmental movement lacks compassion for these groups.

The simplest first step to resolving this misunderstanding would involve environmental organizations consciously integrating people of color, women, and nontraditional groups into their boards and staffs in meaningful ways. By empowering traditionally disenfranchised groups and encouraging their participation within the debate, conservationists can gain critical perspectives on environmental concerns within those communities. Conservation values may then be more palatable to these communities when they are given a voice in the discussion and are assured that their interests and livelihoods will not be sacrificed.

Max Oelschlaeger discusses the need for depolarizing approaches in his thoughtful book, *Caring for Creation*. In his words, ". . . the pervasive idea that there are 'green saviors' who occupy the environmental high ground and 'evil exploiters' who are rapaciously abusing nature is not useful. . . . This dichotomization does not go very far in helping Americans achieve the *solidarity* needed to change behavior through political action and economic restructuring."

3. Build upon decentralized conservation strategies that tap into the "power of place." The past reliance on enacting federal legislation and then litigating to enforce federal standards has indeed served the conservation movement well. However, this approach has serious inherent shortcomings that limit its overall effectiveness as the central long-term tool for promoting conservation. As much as the American public prefers a better environment, the environmental movement makes a serious mistake when the approaches for realizing this goal seem to require increasingly centralized decision making and increasingly bureaucratic enforcement.

Conservation built primarily upon the enduring wisdom of Congress and the federal judiciary is indeed a house built upon shifting sands. It is essential to build a more stable and secure foundation at the state and local levels and by building conservation values into the market. It is also essential to recognize that regulation alone is insufficient to control rapidly growing populations and rampant consumerism. To address these issues requires a personal commitment to the environment from individuals, which cannot be effectively inspired through regulation. The decentralization of decision making can, in some cases, encourage that commitment when individuals are empowered to shape their own community's destiny and their own quality of life.

A New Vision for Conservation in the Twenty-First Century

All over the world and in all sectors of society, we see that prescriptive and centralized institutions are on the decline. Whether in Russia or Canada or in the U.S. welfare system, control over resources is increasingly being decentralized to local management, and even on Capitol Hill, the seat of centralized power in the United States, there is concrete interest in devolving decision-making authority over a great diversity of issues to state and local levels. And in the Supreme Court, interest in a more devolved federalism is even stronger.

Donella Meadows, in a *Grist* magazine article, observed an increasing proliferation of decentralized institutions and organizations that have become ubiquitous throughout the globe. She argues that many of these organizations have positive impacts on the participants compared to traditional, centralized fiat systems, operating under principles of shared purpose and networks of equals. She quotes Dee Hock, founder of Visa: "At bottom, desire to command and control is a deadly compulsion to rob self and others of the joys of living. Is it any wonder that a society that worships the primacy of measurement, prediction, and control should result in the destruction of the environment, maldistribution of wealth and power, mass destruction of species, the Holocaust, the hydrogen bomb, and countless other horrors?"

At the base of this drive to decentralize lies the basic human need for a more meaningful connection between society and the environment, a desire for community to have its own distinctive and unique flavor, not subject to the homogenizing effects of centralization. In this vein, it is interesting to note the extraordinary disconnect between most literature about nature and conservation action. Acclaimed "nature writer" Barry Lopez writes: "The real topic of nature writing, I think, is not nature but the evolving structure of communities from which nature has been removed, often as a consequence of modern economic development. It is writing concerned, further, with the biological and spiritual fate of those communities." Lopez continues: "To include nature in our stories is to return to an older form of human awareness in which nature is not scenery, not a warehouse of natural resources, not real estate, not a possession, but a continuation of community."

North America's great nature writers—from Thoreau to Wendell Berry, Wallace Stegner, Ed Abby, Terry Tempest Williams, Barry Lopez, and Ellen Meloy—all place great emphasis on "a sense of place." However, conservation action in North America has notably overlooked the power of place. Involvement at the community level allows conservation efforts to tap into a powerful force, the commitment and shared identity of people who live close to the landscape. The love of a particular landscape—one's home—can unite even the most disparate ideologies, allowing them to forge their own, widely supported, tailored solutions to environmental problems.

Dan Kemmis, former speaker of the Montana House of Representatives and former mayor of Missoula, Montana, aptly describes this genius of place and discusses approaches for bridging the gap between the power of place and conservation action: ". . . there is much to be gained by allowing ourselves to imagine a structure of sovereignty that is not overborne by the deadening hand of nationhood, but is instead vitalized by the urge of real people to live well together in real places." This genius of place is a powerful, and generally overlooked, tool to integrate local values with a broader desire for conservation. The Sonoran Institute has seen extraordinary results when local residents participate in dialogue about local values, a shared vision for the future of the community, and the actions necessary to realize this vision. This process, undertaken in places such as Red Lodge, Montana, Bluff, Utah, and the San Rafael Valley, Arizona, demonstrates that even in rural regions marked by antienvironmental rhetoric, local residents can produce enduring conservation victories. Conservation, then, becomes not just a passing crisis or a federal mandate, but a way of life inherent in community. This love of place can be used to gain broad, bipartisan support for effective conservation of the landscape.

In my own work, I have found this genius loci, or genius of place, as Dan Kemmis describes it, a tremendous force in bringing disparate groups together to resolve seemingly intractable conservation issues. An effort initiated in 1994 by residents of southeastern Arizona's San Rafael Valley illustrates both the power of place and the importance of a compelling vision to promote enduring conservation. The San Rafael Valley is a relatively pristine valley of gently sloping ridges and rolling grasslands. One of the few perennial grassland streams left in Arizona, the Santa Cruz River, runs through the heart of the valley. The rich and abundant grasslands support nearly a dozen working cattle ranches. These lands are one of the few remaining stretches of unbroken shortgrass prairie left in the Southwest, providing habitat to numerous species that have been driven out of similar environments in Arizona. San Rafael Valley residents were troubled by the rapid growth and change adjacent to this rural, ranching valley on the Arizona-Sonora border. Residents concluded that without local leadership, the valley would be transformed from ranching to rural subdivisions, and they invited the Sonoran Institute to help them develop a vision for the future of the valley and an action plan for preserving both the rich natural values and the sense of community. The result has been creation of the San Rafael Valley Land Trust, which, operating for more than four years, now holds nearly 800 acres in eight separate conservation easements. The permanent protection of the ecologically important 22,000-acre San Rafael Ranch, headwaters of the Santa Cruz River, has also been orchestrated with local leadership through an innovative partnership with The Nature Conservancy and the Arizona State Parks Department. By taking the lead in defining a vision for the future of the valley, residents preserved not only the natural resources and endangered species, but also the sense of community and ranching traditions.

Conclusion

Increased attention to these three principles—articulating a compelling conservation vision, demonstrating compassion for both people and nature, and building upon decentralized conservation strategies—has enormous potential that has largely remained untapped by the conservation movement. To bring conservation action to the next level of acceptance and effectiveness, we must place greater emphasis on what Don Snow calls "coalitions of the unlike." When people with different experiences, worldviews, and backgrounds come together to work out a solution to a land use or an environmental problem, increased creativity can produce results that are often more effective, enduring, and equitable than could be achieved when acting in isolation. This synergy can generate options that were never considered before, placing them on the table for discussion and analysis. This collaboration also affects the behavior of the different stakeholders, who suddenly become responsible for their own actions and ideologies to the rest of the community represented at the table. The experience of listening to alternate perspectives while moving toward a mutual goal also promotes trust. This, in turn, allows future collaborative conservation efforts to proceed with greater confidence and adeptness.

Collaboration, used appropriately, can also lead to the creation of a lasting stewardship ethic within communities, which is generated from individuals within the community attempting to live sustainably in landscapes that they love and to which they feel a personal commitment. If such an ethic becomes ingrained, we may at last be able to bring out the best in our communities, rather than preventing the worst excesses and abuses. As Donella Meadows observes in *The Local Politics of Global Sustainability,* "The trouble is, the sustainable world generally offered by environmentalists is based on restriction, prohibition, regulation and sacrifice. . . . Hardly anyone seems to envision a sustainable world that would be nice to live in." If we can overcome this challenge, and appeal to the imagination of humanity to live well in the world, ' then we will truly have a vision that is inspiring, lasting, and, above all, sustainable.

REFERENCES

Dowie, M., *Losing Ground: American Environmentalism at the Close of the Twentieth Century.* Cambridge, Mass.: MIT Press, 1997.

Kemmis, D. "Learning to Think Like a Region." *The Rocky Mountain West's Changing Landscape* 1:1, Summer 1999, 28–32.

Lopez, B., "We Are Shaped by the Sound of Wind, the Slant of Sunshine." *High Country News* 30:17, September 1998.

Meadows, D., "It's Everywhere You Want to Be." *Grist.* December 29, 1999. www.gristmagazine.com.

Oelschlaeger, M., *Caring for Creation: An Ecumenical Approach to the Environmental Crisis.* New Haven: Yale University Press, 1994.

Prugh, T., R. Constanza, and H. Daly, *The Local Politics of Sustainability.* Washington, D.C.: Island Press, 1999.

Crossing the Great Divide: Facing a Shared History in a Multicultural West

Beverly Brown

I have been asked what working people think about the natural resource collaborative partnerships that have arisen across the West. In my experience, most working people rarely think about this particular set of collaborative partnerships at all. There are good reasons for this lack of interest, and they are practical, pragmatic, and smart. Lower-income working people take an enormous personal risk to participate in controversial dialogue about public lands or waters, and chances are small for workers' issues to prevail. The good news is that the collaborative process—the snowball and self-replicable process itself—has created unexpected opportunities for a reevaluation of the relationship between natural resources and realistic understanding of communities in the contemporary rural West.

Across the West, the social landscape shifted several times over the last two centuries. Almost a third of the West was Native, Spanish, and Mexican for most of a century before the *Mayflower* landed. In the 1870s, my county was one-third Chinese, and Mexican muleskinners brought in supplies for the multicultural mix of gold miners. After ninety years of almost exclusively European-American homogeneity, my rural community today is becoming exponentially more culturally diverse and linguistically rich.

Some years ago, in a meeting at Cornell University, I sat at a table with eight white professors and Indian Studies/Mohawk activist Ron LaFrance, planning a brown-bag seminar (I was the secretary/participant). Ron proposed a discussion by some folks about multiethnic issues in the north country of New York State. The professors said, "But there isn't a problem up there, it's ninety-eight percent white." Ron and I looked at each other, paused, and said simultaneously, "That's the problem!" In a densely multiethnic state, the color barrier was almost absolute at the urban city limit. Except for Native American nations, the informal system of segregation at the rural boundary was far more extreme than what I had grown up with as a European American in the mountain agricultural West.

The essays in this volume reflect a dynamic and growing collaborative movement that, while hardly perfect, has brought diverse interests together to hammer out creative approaches to what have seemed intractable problems. And if not every issue is appropriate to work with across conflicting philosophies, the progress in generating new ways of thinking has been a substantive contribution in itself. As the cultural profile of the rural West rapidly diversifies, towns with 20 to 50 percent Latino, Asian, African American, and/or Native American populations are becoming common. In the collaborative movement, the landowner, and/or educated, and/or business people who make up the movement's participants are fine folks with fire in their bellies and are making an admirable civic contribution. As such, I can testify that it is awkward to perceive oneself as conducting a grand experiment within a kind of closed shop: English required, college-level reading ability mandatory, comfort with public speaking needed, and knowing the generally acknowledged procedural rules of meetings a must. The ability to schmooze as social peers with other participants, including government authorities, is a definite advantage; and, most of all, the ability to participate without risking one's fundamental livelihood is almost always assumed.

Similar to the professors mentioned earlier, we usually accept that the collaborative partnership movement is about who we expect to participate: small (and sometimes large) businesses, economic development professionals and local officials, agencies, and nongovernmental organization activists on governance issues (environmentalists included)—almost exclusively involving a white, well-educated minority. From another perspective, of course, something is awry with this expectation, in which the wide diversity of the multicultural workforce is not integrated into dialogues that will surely directly affect lives and livelihoods among lower-income workers, not to mention the environment in which all our diverse communities live.

Inclusive participation is one of the great gifts of collaborative partnership idealism—democratic participation, the vision of being a part of the design. But we all live within a history that has built a great divide separating lower-income working people of all ethnic and language backgrounds from the kinds of public participation that characterizes the collaborative partnership movement reflected in this volume. One of the pragmatic principles of collaborative work is to narrowly focus goals among a group that includes only "everyone necessary" to accomplish that goal. Lower-income working people, in today's realities of diminished rights of participation and unstable working conditions, pragmatically choose not to even attempt to come on board. Sadly, this mutual pragmatism deepens the divide from both sides and only postpones the time when ignoring the widening inequality in democratic participation wreaks its own havoc on our western landscape in general. In an on-the-ground context, the lack of inclusion of the people who will be actually implementing natural resource directives can easily lead to implementations that are not sustainable.

A more constructive long-range goal will be the integration of concerns from multicultural workers' communities across the West, including open "spaces" that make information and communication accessible across major languages, while respecting traditional knowledge or knowledge gained through experiences working on the land. To achieve this—to make these long-term goals pragmatic realities themselves—is possible if we make a commitment. After all, one basis of the collaborative partnership movement itself is the incorporation of new perspectives.

Twenty years ago, I was sitting in my car in an African American neighborhood of Roxbury, Massachusetts. Almost literally as the stoplight changed, my world shifted, and I suddenly "got it" that although all our European American and African American ethnic histories are radically different, it is existentially "us" in history together as coequals all our extended diverse families, all the collective "us" of aunts and uncles, cousins, and immediate households struggling toward a better future. People who are not from a mainstream U.S. white background will read my epiphany with a wry roll of the eyes, but people from my white culture who have undergone a similar shift from unconsciously defining the norm as white to at least partially "getting it" may understand how even physically startling the realization can be; how we truly didn't "get it" before, even if we intellectually subscribed to the concept.

This stoplight experience was doubly powerful, as not long before, my workplace was involved in a union drive that the employer made inevitable through pure arrogance. Coming as I do from an antiunion family, it was only through research that I came to understand that the hard-fought investment lower-income working people have put into the efforts to improve employment conditions is parallel to the investment of businesspeople in building enterprises.

The Jefferson Center, by which I am employed, has pursued a program to network nontimber forest workers in the Pacific West since 1994, based on a principle of peer-to-peer education—that is, forest workers relating experiences and analyzing forest realities with their peers: other forest workers.

Through this process, which is a collaborative partnership itself across five or more languages and ethnicities, we've learned an immense amount about possibilities of working across cultures. Simultaneous bilingual meetings are "normal," and tri- or quadralingual meetings are not uncommon. My primary experience with the collaborative movement has been with community forestry. Arguably the majority of nontimber forest workers in the West are, considered together, from the Latino, Southeast Asian, and Native American communities, and the remainder are mostly low-income European Americans. These are the people who do the work on the forest floor, planting and thinning trees, watershed restoration, nontimber forest products harvesting, and fire fighting. Even in timberwork in the West, the mills are increasingly showing Spanish and Laotian surnames on the rosters, alongside "Smith" and "Thompson."

When it all boils down, however, and after acknowledging the power of raw ethnic prejudice, one of the fundamental problems shared across all the cultural groups of forest workers is that labor has very few rights in the United States. That is to say, labor has few collective rights. We have expanded recourse in individual contexts: sexual harassment, nondiscrimination grievance procedures, for example. However, through a process of dismantling the Wagner Act of the 1930s in favor of employers, and through changing the workplace to part-time, temporary, contract, and often dubious "independent" work, we've undermined the ability of workers to participate in a democratic process as workers.

Rather than being mourned as a lost opportunity, this has been in most mainstream communities praised as the right of individuals to join the advocacy or community development group of individual choice. Can rural resource workers actually exercise the right to influence fundamental questions about community economics and ecosystems as an individual public participant without experiencing retaliation as workers? Although patronage does not have the formal institutional existence in the rural United States that it has in, for instance, Mexico, it is nonetheless an informal reality alive and well in small and medium-sized towns, where dense social networks easily identify workers who are "out of their place."

On a recent trip to Canada, to attend the first Congress of the Canadian Reforestation and Environmental Workers Society, the limited options of U.S. forest workers became apparent. Among Canada, the United States, and Mexico—countries inextricably linked in forest policy—Canadians have the real right to organize (55 percent of union cards signed guarantees a bargaining unit) and be protected from retaliation by their employer. Mexico and the United States come far behind our northern neighbors, as workers must either rubber-stamp "yellow-dog" company/political party unions in Mexico or fight what has become an almost pointless struggle to overcome employer preferences in labor law in the United States—a direct cause of only 9 percent of private company employees being unionized in this country. Although not every worker wants to organize, the very existence of the effective right to organize is powerful leverage for a workforce.

Intimidation, subtle and overt, is a reality among the workforce with whom the Jefferson Center cooperates. In the week that I am writing this article, my coworker and I sat in rooms in which Southeast Asian mushroom harvesters stated that almost all their compatriots will avoid taking a route of public action that will make them conspicuous and therefore vulnerable (even reporting thefts to authorities), Latino brush harvesters say that undertaking any kind of activity to improve the situation of forest workers will just bring employer retaliation down on them and their families, and two Canadian tree-planters (one European Canadian, one East Indian) on their way to a community forestry conference partially funded by the Ford Foundation were detained for an hour and verbally harassed at the U.S. border by an INS guard who expressed, among other things, that the Ford Foundation "just funds a lot of commie activities."

Our history extracts its own revenge, whatever our intentions. The great divide between typical community forestry participants and workers spans fundamental differences in the literal "practice" of participation. If the community forestry doors were flung open and worker participation recruited across all ethnic communities, the mutual incomprehension of different norms of participation, or lack of participation, makes the meshing of agendas difficult to say the least. Unless differences in languages and literacy are accommodated, the situation becomes even more unsustainable. The worker representatives who are willing and able to participate in current configurations of community forestry, and are accountable to a home constituency, are still few. They are to be applauded in their willingness to help transform the community forestry dialogue, as they are crossing obstacles that few mainstream participants appreciate. How can their numbers be expanded? Or better yet, how can we expand participation and change the context?

Making Change Happen

The investment of community participants in the partnership, watershed, and other collaborative efforts across the West has generated transformative thinking, initiated pivotal experiments, and generally popped open what looked like a closed debate among industry, agencies, the courts, and environmentalists into a democratic process of discovery. Collective creativity and the opening of new doors are a driving force, and in the long run may be one of a handful of the most important legacies of the partnership movement.

Given all this, there's no reason why we can't change the way we're putting together our overall frame of reference.

Ideas move actions. Collaborative processes have opened a ladder of information and potential routes into the public dialogue. The existing movement has helped generate interest in activities that are pushed by, react to, or simply are energized by the idea of the participatory process of the natural resources partnership/collaborative movement. Because of this momentum, I have watched as forest workers, through peer-education activities, put together the foundations of multicultural, multilingual collaborative networks parallel to the mainstream English-only collaborative process. Workers designed these collaborations to address the realities of working people's lives, both respecting ethnic identity and as major stakeholders in forest and forest watershed issues. Although beginning as parallel processes, they are on a converging trajectory with mainstream community forestry efforts, preparing not only "alternate proposals" but also an alternate and more comprehensive vision. Forward-looking mainstream groups have opened doors for the worker-activists. An excellent example of this worker-generated process is the emerging Alliance of Forest Workers and Harvesters/*Alianza Trabajadores y Cosechadores Forestales.*

Predictably, funding and support are more scarce on the worker side of the great divide. The concept of "environmental justice" has yet to reach deeply into the world of natural resources. A shift in understanding by environmental and community-development-focused foundations could make viable more equitable collaborative initiatives. Currently, only a few small social justice foundations and private donors are funding these efforts. The Ford Foundation has so far been the one large foundation to directly recognize the importance of funding these worker-based community forestry efforts.

One of the great strengths of collaborative partnerships is that who is "right" becomes a fluid intellectual undertaking, allowing new information and perspectives to be considered alongside environmental, industry, and other hard-line political positions. Still, it is who has power that determines the final outcome. As worker-initiated natural resource networks develop, workers will inevitably address how labor law, color barriers, and the punitive social structures that support patronage power curb the ability of natural resource workers to publicly participate in determining the future of healthy communities and healthy ecosystems.

From environmentalists and others, my coworkers and I have repeatedly heard the criticism that the concerns of a diverse cultural group of forest workers are simply "labor disputes," somehow surgically detached from working people's concerns for environment and community. The naivete of this allegation makes manifest the depth of the great divide in our society, when, for instance, environmental groups seem oblivious to a whole new realm of support if only resource workers—other than the minority clustered around the company owners and TV cameras—were free to speak their minds. Community forestry advocates, whether from the right wing or the left wing of the movement, seem complacent that the H2 guestworker ("new Bracero") programs championed by community forestry "supporters" such as Senator Larry Craig (R-ID) can eventually undermine efforts by community forestry advocates to create "family wage jobs." Instituting an "indentured servitude" foreign worker program will actually undermine any attempts by all workers to gain fundamental human and democratic rights on the job or in society.

The possibilities for alliances are implicit, but problems that share root causes will inevitably be interpreted differently. Whereas mainstream groups fight for their version of community inclusion or courtroom victory in the bureaucratic process of ecosystems management, diverse cultural groups of workers are often fighting for fundamental human and civil rights, a battle for basic economic, social, and environmental justice in which the issues are inextricably linked. What we need is to see the interlinking of these issues and how that plays out in the participation or nonparticipation in collaborative processes.

In the community forestry movement, there is already a small but pivotal group of forest workers from the Lao, Cambodian, Mien, Latino, Native American, East Indian, and low-income European American communities who are willing to make their voices heard in the mainstream. A smaller but crucial group of mainstream participants have begun to travel—figuratively and literally—to the diverse cultural communities of forest workers, and a few of them have made the frame-of-reference shift that makes ethnically diverse worker issues not an "add-on," but part of the integral vision of participation.

The absence of ethnically diverse scholars in natural-resource-related studies is an especially serious blow to natural resource worker-activists. It is the rare and exceptional European American researcher who can grasp the natural resource subject matter and also accurately represent, for example, the complex labor, cultural, and environmental issues of an immigrant community working in the woods. (The same is true for middle-class researchers approaching low-income European American forest workers.) Yet it is not the exceptional, but the typical university researcher who usually undertakes such studies, and the result is all too often that the less-known and lower-status cultural group predictably becomes the "problem." Because these studies are circulated among decision makers, and peer-reviewed by researchers who also lack an understanding of cultural contexts and interpretive subtleties, information goes out that inaccurately represents highly complex situations.

Getting young people from low-income natural-resource-worker and ethnically diverse communities into college programs dealing with natural resources—including environmental sociology—is a critical link in consolidating and legitimizing the participation of culturally diverse low-income-worker participation in collaborative processes.

And finally, ideas must lead to action. Pilot projects on the ground are the only way to bring all these concerns together. Projects initiated by or inherently incorporating workers as active agents in the dialogue, design, and implementation (rather than the tail-end recipients of "jobs," as is typical today) bring the process of diverse partners, ideas, and experimentation to fruition. Many partnership groups—for example, watershed councils near corporate agriculture and urban areas—are dominated by powerful established interests, a circumstance that discourages worker participation. Extra attention to funding and support will be necessary from foundations, agencies, universities, nongovernment organizations (NGOs), and concerned citizens to assure that worker-inclusive pilot projects become reality.

The short and competitive timeline of policy making pushes us too often to practice *realpolitik*. We pursue the shortest path to apparent progress, leaving out everyone who is too slow to the draw, ultimately undermining our own design of action. The longer road and broader vision will take more time and challenge us all in reconsidering our respective definitions of "community." The European American majority across the Pacific West will be a historical footnote in little more than a century. Most of the descendants of both Anglo and non-Anglo communities will still be workers who do not reach the high ranks of society, but who will share the consequences of natural resource policy. What kind of future do we intend to build?

Collaborative Conservation: Peace or Pacification? The View from Los Ojos

Maria Varela

"Sin Tierra No Hay Justicia: Sin Justicia No Hay Paz"—"Without land there is no justice: without justice there is no peace"—graces a banner hung on the marbled walls of the New Mexico State Capitol Rotunda during the 1996 legislative session. In front of the banner, weavers arrange multicolored piles of wool, a spinning wheel, and a loom. There is a table of wild-crafted herbs, a corner where a drum maker works, and a display of micacious clay pottery. A woodcarver, oblivious to the throngs of politicians, lobbyists, and others attendant to the political process, remains absorbed in bringing a face out of gnarly wood. The materials used by these artisans originated in the national forests of northern New Mexico. *Poquiteros* (small-scale sheep and cattle growers), weavers, woodworkers, potters, and *curanderas* (healers) came together to make a gentle but powerful statement against environmental organizations seeking to close public lands to grazing and timber harvesting, as well as to capture ancestral village and tribal waters for instream flow.

The weavers, reluctant warriors in this latest struggle by villagers to retain centuries-old pastoral cultures, would prefer to be back in their workshop in the northern New Mexico village of Los Ojos, educating visitors through displays of beautifully woven rugs and tapestries. There, it is more comfortable to talk about how their enterprise, Tierra Wools, and its parent organization Ganados del Valle (a nonprofit with the mission of creating sustainable economies from cultural, agricultural, and natural resources), rescued the rare *Churro* sheep breed from near extinction and revitalized the centuries-old Rio Grande weaving tradition. But the weavers recognized that in this continual struggle, artistic expression alone will not safeguard ancestral land and water rights.

Several hours south of Los Ojos, from pricey refurbished adobe and Victorian offices in Santa Fe, New Mexican environmentalists have fired off a nearly decade-long barrage of lawsuits, seeking domain over grazing and timber resources on public lands under the Endangered Species Act. To many northern New Mexicans of color, these barrages are the latest in a conquest that began in 1848 with the signing of the Treaty of Guadalupe Hidalgo and have not stopped since. The battlefield in this latest incarnation of conquest is in the courts and in the streets. In the fall of 1995, environmentalists shut down the Carson and Santa Fe National Forests to all timber harvesting (including fuelwood, which villagers depended on for cooking and heating) because of the presence of spotted owl habitat. (The U.S. Forest Service spent almost $2 million during the early 1990s attempting and failing to find the owls in either national forest.) Unable to afford legal council to respond to the injunction, villagers launched protests. Environmentalists were hung in effigy in October. Two months later, a candlelight prayer vigil was organized in Santa Fe, bringing out more than four hundred people from northern and southern New Mexico. Unwittingly, the environmentalist's lawsuits had done what had never before been achieved in New Mexico's history: a collaboration between Anglo ranchers and loggers from the south with the traditional Hispano *Poquiteros* and fuelwood harvesters from the north.

There is a long history leading up to the candlelight procession on that snowy December night. It began in the late 1890s, when U.S. and European resource barons clear-cut the bountiful forests that stretched almost uninterrupted from Santa Fe County north through Colorado, damaging the snow-holding capacity of upstream timber stands, leaving downstream agricultural villages to struggle with silt-laden irrigation ditches and low or no water during dry years. Then in 1905, Theodore Roosevelt, ironically a hero of the conservation movement, commandeered the common lands of villages and pueblos (supposedly guaranteed by the Treaty of Guadalupe Hidalgo) into the Carson and Santa Fe National Forests. After that, livestock barons purchased grazing allotments on the forests and sent truckloads of cattle from Texas, Oklahoma, and eastern Colorado, seeking the plentiful grasses that once sustained village and pueblo flocks and herds. Local families, still on a mixed barter and cash subsistence economy, found it difficult to compete with the grazing barons for these grazing allotments.

As bottomland villages were severed from upland grazing ranges, valley floor pastures became degraded and lost their capacity to sustain local communities. Before conquest, *Poquiteros* moved livestock from lowland to upland ranges and back again. As a result of this intensive but well-timed grazing system (with origins on the Iberian Peninsula), forage maintained its vigor and grew so high that elders recount how "we'd have to be up on horseback in order to find the lambs or calves." Losing uplands grazing meant reducing herds and flocks to sizes that were not economically feasible. But because livestock were handed down from generation to generation, buffering families from hunger as well as symbolizing connections to ancestors (livestock were often used as dowry or gifts), many villagers clung tenaciously to their pastoral culture, which once fed family, village, and region while renewing the environment.

Those common lands not commandeered into the National Forest system were seized for back taxes, stolen by Santa Fe–based land speculators, or sold by village opportunists without consent from the rest of the residents. All of these acts broke the Treaty of Guadalupe Hidalgo, which promised that "property of every kind, now belonging to Mexicans . . . shall be inviolably respected . . . and all Mexicans shall enjoy . . . guarantees equally ample as if the same belonged to citizens of the United States." As land eroded out of village and pueblo control, generations of knowledge about how to live on and with that land began to unravel. Although rarely recognized as such, this knowledge is primary environmental science. How to treat the grasses, the waters, the forests, and the soil in distinct ecosystems is a science steadily built over time through a process of trial and error. As this knowledge was ignored by outside forces and, in some cases, inside *opportunistas,* environmental degradation, poverty, and its bitter social fruits soon followed.

Counties north of Santa Fe, in both New Mexico and southern Colorado, are designated by the U.S. Department of Agriculture as "Persistent Low Income Counties." This means that since the census started tracking income levels (1939), these counties have remained in the bottom quintile of all U.S. nonmetro counties. Broken treaties and the industrialization of livestock production and timber harvesting over the last 150 years left villagers to eke out sustenance from several sources: sales of livestock, fuelwood, hay, flagstone, vigas, pinon nuts, and traditional arts and crafts. In the past, these revenues were supplemented by the most able-bodied in the family moving into the migrant farmworker stream, leaving the "ranch" to the elderly or to women with small children. Today, local villagers still take seasonal jobs or commute to urban centers, working low-paying, full-time jobs while maintaining "the ranch" back home. The goal is to ensure a continuation of *la herencia,* or the handing down of family lands to the next generation, as well as to protect oneself from the deep poverty often experienced by those elders who have lost their land.

Conquest Reconstructed

Environmentalists, many of them new transplants to New Mexico, look at the ravages left by the industrialization of public lands and, perhaps to their credit, have decided to fix it. But it is unfathomable why environmentalists would not look to the people who have lived for hundreds of years in agropastoral communities, which buffer public lands, as the first to be consulted and as peers in this effort. After all, Hispano villagers and Native Americans have lifelong knowledge of these lands and have fought extractive industries and the U.S. Forest Service long before the modern environmental movement. By rendering people of color invisible, or vilifying them as "violent" or "tools of livestock and lumber transnational corporations," many environmentalists have, in their historical and cultural illiteracy, assumed the cloak of conqueror.

Unexamined by environmentalists is the violence of poverty. Persistent poverty and economic dependency have resulted from a century-and-a-half process where national and international capital forces sought to gain wealth and "make more productive" the seemingly "idle" or underutilized natural resources, land, and labor of western communities of color. As industrialists had their way with these resources, poverty and environmental degradation became a structural part of much of the rural West. It is not possible to repair the environment without repairing the inequities produced by this violent history.

The single biggest threat to a sustainable environment are the so-called "New West economic growth strategies," rampant throughout the region. This latest colonizing wave comes on skis, golf carts, river rafts, and Jeeps. Many environmentalists view recreation and tourism as "walking more gently on the land," a therapeutic antidote to mining, logging, and grazing. Yet the growing tourism and recreation economy in the West has resulted in increased air and water pollution, degradation of scenic resources, and loss of agricultural lands (and therefore wildlife habitat) to subdivisions and resorts. Proponents of a more environmentally beneficial economy generally dismiss grazing as economically unimportant to the West, while ironically the demand for naturally grown meats and other natural agricultural products continues to increase annually. Others claim that tourism/recreational service jobs pay better, require more skills and higher educational levels, and have more potential for advancement. Unexamined is who benefits from the higher-paying jobs and who gets the menial jobs. Virtually no published environmentalist has examined the impact of racism and economic inequity in a heavily dependent tourist/recreational economy.

A case in point is the growth of the tourism/recreational economy in Taos County, New Mexico. In the 1950s, local people were told that expanding tourism would create jobs and improve incomes. Since the 1960s, gross revenues in Taos County have risen commensurably with the expansion of recreation and attendant retailing, restaurant, and service businesses. But even when economic activity doubles, as it did between 1980 and 1990, poverty statistics remain virtually unchanged. In 1980, the official poverty rate for Taos County was 25.7 percent. In 1990, it was 25.4 percent.

Clearly, the increased flow of revenues through the tourism/recreational economy of Taos had little effect on material poverty and has actually exacerbated ancillary poverty, including loss of ancestral lands and water, loss of resilience in family-income-generating activities, increased taxes, and the erosion of community cohesion produced by growing racism and economic inequality. In the fall of 1999, a recent transplant to Taos published a feature article in the *Sunday New York Times* where she recounted incidents of perceived hostility toward her by young Hispano males, which resulted in her decision to purchase a handgun for protection. She wistfully recounted her reasons for coming to this breathtakingly beautiful mountain valley from her previous urban residence and expressed anger at Taos's "dirty little secret" of hostile and violent locals.

Reconnecting Culture, Economics, and the Environment

In contrast to the neocolonialist economies and increasing social stratification in the New West are the efforts of those indigenous to the region who seek to build sustainable economies by reconnecting the best traditional cultural practices with modern production and marketing strategies. These groups focus on adding value to cultural, natural, and agricultural resources. Educational and cultural tourism are part of these economic development strategies, offering the opportunity to educate mainstream America and to create niche markets. These strategies strive to benefit those with limited economic status and educational levels and whose gender, race, and ethnicity have left them economically marginal. Such strategies require long-term, patient investment to underwrite human-capacity-building and to meet the research and development needs of innovative, local enterprises.

Ganados del Valle, for example, spent seven years (1984–1991) supporting the technical assistance and training needs of Tierra Wools weavers (operational costs were covered by sales). In 1992, Tierra Wools annual sales reached nearly $350,000, and in 1997, the business was spun off to its weaver-owners. Ganados went on to create four more enterprises from 1990 to 1996, which market naturally grown, local meat and produce, regional arts, crafts, and home decor items. More than 150 artisans and agriculturists in the region have been assisted with bringing their products to market, and 50 new jobs, many in management, were created locally. A work-based academic program was designed to professionalize staff, nearly all of whom are women with high school diplomas. A small loan fund helps artisans and growers to be more productive, and a scholarship fund supports those wanting to complete college degrees.

From Conflict to Collaboration, and Back Again

Despite these accomplishments, the efforts of Ganados were constantly laced with conflict over use of resources. Whether protesting the transfer of agricultural water rights to a proposed ski resort, supporting local landowners in a title dispute with real estate developers, or supporting forest-dependent communities in a lawsuit against the Forest Service for discrimination in the allocation of timber and fuelwood sales, Ganados quickly discovered that one of the major barriers to a sustainable economy was control of natural resources.

The most difficult struggle for Ganados was over access to summer grazing for members' flocks. Because the organization returned to the tradition of cooperative grazing (where flocks are pooled and flock owners share the costs of a shepherd), small-scale growers were able to bring their flock numbers up to an economically viable level. This put more pressure on Ganados to locate summer pastures for the growing flock, which was needed to supply the annually increasing demand of the weavers for high-quality wool, especially from *Churro* stock. In the mid-1980s, Ganados began discussions with New Mexico Game and Fish about the possibility of grazing one of the two state wildlife areas in the valley. Traditionally, these areas had been grazed by Anglo cattle growers. Ganados, in consultation with the local office of the U.S. Soil Conservation Service (SCS), offered the flock as an intensive grazing management tool to improve the quality and quantity of forage on the refuge, while eradicating what SCS identified as invasive shrubs and weeds with little or no value for wildlife. After nearly two years of discussion, New Mexico Game and Fish settled the matter by convening a task force. Its first order of business was to put an indefinite moratorium on all grazing in wildlife areas.

In the summer of 1989, after a fruitless search for grazing lands for the cooperative flock, the Jicarilla Apache nation agreed to let the sheep graze on its lands. However, several weeks into the summer, the tribe's attorney notified Ganados that the lease was suspended because of pending litigation with Game and Fish, which owned the wildlife area bordering the Apache's land. Bringing the flocks back to home pastures in July, before winter feed crops were fully grown and harvestable, would have forced growers to sell their flocks, including the *Churro*, because they could not afford to buy winter feed. By this time, some growers were in their fourth year of a seven-year breed-back cycle required to return the *Churro* to its original characteristics. After a failed appeal to Game and Fish for emergency grazing, growers moved their flocks on a moonlit July night onto the adjacent wildlife area in an act of civil disobedience.

As soon as news of the "sheep-in" hit the media, major New Mexico environmental and hunting organizations were quick to condemn it. (The wildlife areas are used primarily by Game and Fish as premium hunting areas for trophy elk.) The shrillness of these attacks took Ganados by surprise, as we had pursued environmental goals over many years on issues ranging from predator control to sustainable agriculture to fighting inappropriate subdivisions. Ganados leaders are practicing environmentalists and assumed that the card-carrying environmentalists valued our efforts to protect land, water, wildlife, and local cultures.

Realizing that the strong coalition between hunters and environmentalists would be difficult to defuse, Ganados went to U.S. Senator Jeff Bingaham, considered the state's "environmental" legislator, to request assistance in convening a professionally mediated retreat. Twenty environmental leaders were invited to the table. Five agreed to come. Ganados found resources to house the group for three days and retain the mediators. Initially, these meetings seemed to defuse mutual hostilities. Ganados leaders learned to respect the dedication of the environmentalists. Environmentalists appeared genuinely moved when, after a lunch prepared by the weavers, the women made presentations on how important Tierra Wools and the members' flocks were to their lives. One weaver said sweetly and candidly, "and if we had to put the sheep on the wildlife area again and go to jail, we would do it. It means that much to our families and our community."

Toward the end of the retreat, after sorting through commonalties and differences, the group resolved to go beyond "just talking." Action around a mutually agreed-upon project would create genuine collaboration that could create lasting alliances. A land-purchasing project in the Chama Valley was conceived, which would meet the environmentalists' desire to buffer development in prime agricultural areas while meeting Ganados's need for grazing to continue its sustainable development strategy.

But instead of creating lasting alliances, this joint project eventually failed, revealing some core problems with collaborative approaches. Ganados learned that the environmentalists themselves are not united. Their different membership bases and corporate cultures often result in different approaches to environmental problems. Those who came to the table felt they had little influence over those who did not. Those who did not come to the table felt that they didn't need to. They could achieve their environmental goals through the courts and legislatures. Those who came to the table were unwilling to stand up to other environmentalists to ask them to reconsider their tactics. Their continued collaboration with Ganados opened them up to criticism from the nonparticipating groups. When those who worked for national organizations found themselves growing closer to Ganados's point of view, they often ended up at odds with their superiors and membership. One member of the group dropped out because his board would no longer underwrite his work on the joint land purchase project.

In the end, it was Ganados who really needed the joint land project in order to continue its sustainable development strategies. Those environmentalists who participated wanted to demonstrate that they and their organizations were not insensitive to the economic and cultural needs of New Mexico's communities of color. But this was a moral gesture, not a need. When parties in collaboration are not there because of equal needs, the effort depends on the "charity" of those wanting to make a moral gesture. Charity does not result in justice nor does it reverse the ravages of racism on either side. Without economic justice and authentic cross-cultural learning, collaboration is an exercise in the pacification of rural populations.

Until environmentalists come to need the contributions and continued existence of agricultural cultures to stay on the land, supported by profitable and sustainable production and marketing strategies, their participation in the collaborative process is more of a moral stakeholder without actual investment. At the same time, in order to stay on the land, agricultural cultures must move beyond colonial economies and create new economic opportunities, adding maximum value to sustainably grown and harvested resources. In this process, communities of color will need the environmental community to advocate for more environmentally and economically beneficial policies for rural America.

Collaboration can provide the opportunity for the kind of cross-cultural communication that is necessary to address social, economic, and environmental problems in the West. But unless the issues of race, class, and culture are faced head-on, I question whether collaboration can make a dent in deeply held ethnocentrisms, rooted in still deeper historical legacies. Breakthroughs are possible, but only if we can gather the courage to risk stepping outside our colonized worldviews.

Finding Science's Voice in the Forest

Jim Burchfield

At a recent meeting of leaders of several western National Forests, a lengthy discussion on collaborative planning turned to the timing of public participation in the planning process. One of the Forest Supervisors with a reputation for an open, inclusive administrative style mentioned that her National Forest would complete a thorough description of natural resource conditions prior to involving people external to the agency. The description would allow people to understand what was happening within the forest. It wouldn't affect people's engagement in the planning process, since, after all, the analysis of the management situation was "just data."

Just data. Wouldn't it be lovely to think so? The opinion of the forest supervisor mirrors the worldview of many environmental researchers, that scientific information provides a particularly meaningful portrayal of our place in the world. A powerful myth has captured many thoughtful and well-meaning professionals, that there is such a thing as "good" science, and it will be our true guide on the path to righteousness. Yet when we peel away the layers of any science report, we discover otherwise, that science is more art than truth, created by people who operate, like all of us, in a conscious and unconscious universe. Guided by "inner voices," as microbiologist Robert Pollack calls them, researchers are inspired to discovery. But the great gift of our minds carries a long recognized cost: We mix ourselves in our work. Hidden within each table of measurements are the values of researchers, who have constructed unseen assumptions and numerous judgments over what may be worthy of display. Data are not just data. Information is always accompanied by interpretation.

Science is a powerful ally. It allows us to explain and generalize our experiences, opens doors to understanding, gives us tools and devices for survival and comfort, and instills confidence that the bridge will not collapse when we drive across it. Representations of natural resources conditions and trends using protocols of science will remain essential to understanding cause-effect relationships and the consequences of policy decisions. But we must be very cautious when we turn over descriptions of reality to any group, no matter what its self-proclaimed piety.

As we enter into new eras of awareness and more complex planning environments, we are forced to look at the contributors to natural resource conservation and management in a new light, ever mindful of assignments of privilege. Environmental researchers are not oracles. They are more like artists, assembling and mixing the colors of an awesome, complex, and dynamic story into a coherent picture. Data may be the paint on their pallets that creates a picture, but inevitably, what we see is the artist's rendition. Decisions are made at multiple steps along the way: on relevant facts; on the boundaries between what we know and don't know; and on what we care about. As long as we recognize the artistic side of science and the inherent uncertainty surrounding its findings, we can move forward, building on our collective understanding and talents. Yet if we interpret science as something other than this artful endeavor, if we sense that it offers true direction, that its elegance takes priority over other forms of knowing, we can be misled, disappointed, and ultimately suspicious of our most trusted partner in solving human problems. Finding the appropriate role for science in our efforts to make policy decisions remains one of natural resource management's greatest challenges.

The Guiding Light

Science is never certain; its claims are continuously and appropriately contested. Either new discoveries replace old hypotheses or investigations are challenged on methodological grounds or measurement error. Natural systems are so complex and dynamic that explanatory powers quickly diminish with increasing size or expanded time frames. More important, the division of science into disciplines (a necessary specialization to attain sufficient focus and depth) makes integration among simultaneous phenomena exceedingly difficult. Different disciplines use different methodologies, different languages, and different standards for scaling thresholds of concern. The resulting confusion of methods and disputable results leads to a general frustration with what may be valid or reliable information. Moreover, since research strives toward "external validity," or the capacity to generalize findings to other situations, the reach of research can easily exceed its grasp.

Researchers understand, of course, the provisional nature of their discoveries. Our best environmental researchers, such as conservation biologist Gary Meffe, are forced to admit that "the only thing certain about ecological systems is their uncertainty." This does not, however, restrain them from attempting to leverage influence in a decision's outcome. Adoption of their work affirms their labors and acknowledges them as valued contributors to the community. Yet what is all too frequently ignored is the personal side of researchers. Like all other participants in a decision process, they possess both interests and biases. A most difficult challenge is the recognition that researchers are not just searching for truth, but they are engaged in a process of negotiation. Sheila Jasanoff observes this in her study of regulatory processes for environmental quality. It is the duty of decision makers to struggle with the contributions of the science community without being captured by one of its specific legions. Since we recognize there are only partial truths, Jasanoff recommends that we must be satisfied with a "serviceable truth," one that does not presuppose certainty, but moves us toward a common goal.

Environmental scientists recognize that their predilections are crucial in choosing variables. Sharachchandra Lele and Richard Norgaard argue that biologists possess an understandable affinity toward biodiversity, but further, they are forced to pay attention to the concerns of their research sponsors. Those that fund research often represent the interests of the powerful (governments, foundations, banks, landowners) over the interests of the weak (rural peoples, subsistence users, the unemployed). With acknowledged ideological frameworks operating within the hearts of researchers and the checkbooks of sponsors, researchers must be especially attuned to limitations of their own work.

Even with thoughtful, careful, and noble researchers, there is always the potential for those who frame the problem to ask the wrong question, especially if they represent a limited worldview. Investigators are forced to reduce a complex system to a handful of variables to conduct analyses, but which variables are used within models remains a subjective choice. Problem identification is the core of planning—collaborative or otherwise—and whoever controls the description of the problem controls the plan itself.

This is why scientific assessments, front-loaded in many public sector planning models, can become problematic to meaningful multiparty participation. If researchers decide what must be measured, they also, by default, are deciding what must be important. When we examine how science is cast in the policy process, what we perceive as science and the authority we grant its conclusions are central considerations.

In Process We Trust

Policy makers must be very careful in awarding special status to science in policy deliberations. Among most participants in multiparty planning processes, scientific information carries an influential weight because of the rigorous standards that guide its reporting (notwithstanding the consistent reminders about the rarefied air that its initiates breathe). The arrival of even modest rigor to characterizations of conditions is often welcome, particularly in localized, collaborative decision-making settings. These groups frequently experience independent testimonials to fears and folk stories, which are passed off as accurate representations of cause-effect relationships. In settings such as these, where access to expertise is limited, scientific information nearly blinds participants with the apparent glow of its logic. However, a principle of the collaborative process is equal access, not only to the expression of interests, but also to interpretations of critical facts, events, and relationships. Scientific information easily passes the entry exam for quality control, since it must, by definition, be information that has been systematically collected, organized, and reviewed. Whether it guides discussions and decisions in a fair and open manner is another question. What is missing all too often is critical reflection on whether the information should be accorded special status or if it is even relevant. By virtue of its descriptive powers, science can accidentally become a privileged member of the group.

Placing science information in a central role in the decision process also creates an illusion of a technical character of problems. If issues are portrayed as mechanical imbalances or situational departures from a fixed standard, then clearly the issue can be resolved by technical adjustments. An assertion of misalignment as the root of problems empowers agencies to be repairmen. Instead of managers or administrators who are intended to represent the broad and often conflicting interests of citizens, the rational professional grabs the most competent engineer to impose a fix—possibly a new dam, a few more bends in the river, or some genetically improved trees. If this doesn't work, the manager presses onward, hiring additional experts, who create an even more impermeable, incomprehensible bastion of specialized knowledge. Instead of a diversity of views that transact to arrive at a provisional, somewhat doubtful step to the future, we allow entire regions to be guided by optimization models. The utility of these models has been questionable at best, with an all too high probability of "garbage in and garbage out." A belief in technical control can impede the realization that we cannot and should not expect to be able to run the whole show.

On the other hand, science capably addresses many questions in a collaborative decision-making setting, and well it should. What it cannot address is the desirability of the conditions described—whether the fire risk is tolerable, whether wildlife interactions are acceptable to landowners, or whether the forest supplies the kind of wood products that meet community needs. These are normative decisions, based on what members of the group perceive is appropriate and affirming for the interests of the community. However, what typically happens is that standards about the appropriateness of conditions are preloaded within science reports. Forests are not "properly functioning" or they are in need of restoration. Restored to what and for whom? Since when are the standards for "proper" set by science? Science can tell us whether we are within or outside of given standards, but not the standards themselves. Standards are judgments made on factors that are socially and political desirable. The admonition that the forest is "out of whack" may be attributed to a determination by "science," when we are actually seeing the values emanating from the researcher.

The outputs of science, maps, computer images, and tables of numbers can often act as bullies on the school playground. These technical wonders, although helpful in deducing the components of specific phenomena, can overpower and even marginalize other interests with the charm of their sophistication. Other interests may have legitimate claims, but are less able to put their concerns into such dazzling formats. It's not as easy to map the nuances of a neighborly exchange of livestock pasturing as it is to display four-color approximations of elk winter range on the very same ranchland. In permitting technically driven characterizations to substitute for explanation, other factors affecting outcomes can be easily discounted. Proponents of views that contrast a computer model's output quickly discover that their arguments receive less weight in policy discussions.

This does not imply, however, that we must abandon science as a tool to support policy development. Science may have a more valuable role as a model of process instead of as a producer of results. The methods of science—observation, experimentation, and verification—could serve as guideposts for rigor in deliberations. As political scientist Erik Albaek thoughtfully observed in his comparison of science and policy making, they both rely on mutually reinforcing procedures. Each is an idealized process with flaws and twists determined by the participating actors, but each allows a logic that gives observers opportunity to levy substantive critiques if procedures are violated. Both decision making and science may be considered artificial, as they are based, in part, on the norms of the cultural and historical context. Yet neither is arbitrary. Each must adhere to recognized boundaries of behavior to allow their functions to be serviceable.

The discoveries of science will continue to contribute to policy, but they could offer more if their applications were rearranged within decision-making procedures. Why isn't scientific information more common at the end, instead of at the beginning, of the planning process? The transactive nature of planning requires not only that we converse about what we should do, but that we also learn from our actions. If collaboration is intended to support learning—and to me, learning is the most vital outcome of collaboration—then managers, citizens, and researchers need to use the best learning tools available. Where is science, then, when monitoring takes place? Where are the public sector scientists after the completion of a timber sale, the closing of a road, the building of a bridge, to observe the effects that they spent too much time describing prior to the project? Why don't we apply science more frequently in evaluation? Did the intervention achieve what it claimed? Why or why not? What can we learn to do better the next time in a similar situation?

Unfortunately, there is no glory in evaluation. Few wish to advertise that the project didn't work, that the model was wrong, that the assumptions didn't hold, that it rained and everything turned to mud. It is much more exciting and publishable to estimate in advance—to advocate—than to debrief. Researchers are all too absent from the postproject phase of the planning process, the one area where they have indisputable winning cards. If this is because sponsors are reluctant to reveal the actual effects of planned actions, then something is amiss. Researchers should not be accountable for the consequences of actions, since they should not recommend actions in the first place. Or have they done so? This is the rub. When it finds itself in the tenuous position of the advance scout, science loses its ability to perform more credible, reflective work. When science determines a solution, it becomes a sponsor, an invested party in the outcome. It threatens the trust we afford to science and undermines it greatest strengths. What happens to science when it is forced take blame for a failed prediction?

In fairness, researchers are not fully at fault for directing policy. The dispiriting collapse of leadership throughout policy-making institutions deserves much of the blame. In a painful, repeated, and embarrassing display of timidity, elected officials and their hired administrators have done everything in their power to avoid making choices. Instead of doing their jobs, they pass the buck to science. The current generation of public sector administrators has been unwilling and unable to activate a fully developed, transparent policy-making process. They have raised the banner of "science" answers to avoid their own responsibilities. Researchers, operating out of a combination of duty, ambition, and hubris, accepted the mantle of leadership, sadly misjudging their own mission and lacking the wherewithal (or the assurance of future funding) to tell their bosses to earn their own paychecks. From the point of view of administrators, it's been a sweet deal. Gone is the accountability derived from making choices. "I was following science" lingers as a convenient excuse.

The emergence of collaborative planning processes has been an attempt by ordinary citizens, in the midst of myriad demands on their time, to try to get things done. When they confront the unwillingness of leaders to cooperate, they do the best they can on their own. But frequently, they stumble over the mighty idols of science, especially big science, with its linear programs, multicolored maps, specialized language, and insular style. Citizens have little ammunition with which to challenge the priests. Sometimes, volunteer groups hire their own scientists, to joust on the field of statistics over characterizations that are commonly tangential to the problem at hand. It would be more helpful to simply reform governance itself, since leaders and responsible institutions remain central to the management and disposition of natural resources.

Conclusion

What can be done about all this? I would first suggest that researchers must become accustomed to explaining the interests behind their work. For what purposes is the research conducted? Who may benefit and in what ways? The explication of researchers' aspirations is a fundamental prerequisite to honest participation in the political process of decision making. We typically ask other stakeholders what they care about. We should ask the same of researchers. Second, it is time to put the "science assessment" parade to bed. For the past decade, public sector land management agencies have been paralyzed by science assessments. Enough already. Instead, we could evaluate the outcomes of specific actions, based on modest objectives, so we might learn while making progress.

Reducing the scale of policy decisions to dimensions accessible to human deliberation could help enormously. This is not to say that basic research on large-scale processes—global climate change, for example—should be abandoned. These investigations are needed to inform us of the context of our decisions. Yet by lowering the expectations of a large body of environmental science to issues occurring at a finer level of resolution, environmental science can take small but powerful steps in solving problems.

If conclusions of research depend on assumptions within the minds of researchers, then why not allow those assumptions to be tested within the mix of ideals and desires of collaborative groups? Let science questions arise from the community, not from the agencies, not from the "data." Allowing only scientists to determine the decision frame threatens the greatest strength of the collaborative movement, its creativity. The effort to revitalize the democratic experiment at the community level can benefit from the skepticism and replication that guides research methods. Allow the approach, not just the findings, to be a fundamental contribution of science.

We must all accept and help inform our leaders that there is no "scientific management." When authorities of any stripe say that there are "science" answers, or that we will base our decisions on "good science," watch out. There is only management that more or less advances certain values. Research can give us insight as to whether a specific action will help or hinder our objectives, but it will be provisional and only partly true. To assume that we know, via science, what we should be doing is destructive to the tradition of science.

Instead of falling back on science, leaders must be clear about the purpose of management. Let us not shroud our values with our most useful tool. Science is not the same as politics. We need both to work through the tough problems of how natural resources can be allocated and used. Science is our partner, not our master.

REFERENCES

Albaek, E., "Between Knowledge and Power: Utilization of Social Science in Public Policy Making." *Policy Sciences* 28, 1995, 79–100.

Jasanoff, S., *The Fifth Branch: Science Advisors as Policymakers.* Cambridge: Harvard University Press, 1990.

Lele, S., and R. Norgaard, "Sustainability and the Scientist's Burden." *Conservation Biology* 10:2, 1996, 354–365.

Pollack, R., *The Missing Moment: How the Unconscious Shapes Modern Science.* Boston: Houghton Mifflin, 1999.

"Salmon Is Coming for My Heart": Hearing All the Voices

Shirley Solomon

I live on the delta of the Skagit River and I work to restore salmon. My home is an old farmstead known locally as the Stroebel Place. Now highly productive industrial farmland, this was tidal marshland in another time, well used by native people and by salmon. The Squinamish, who did not survive the successive waves of white contact and the changes of that period, were once here. And so were the salmon, from the time they colonized these postglacial waters, some ten thousand years ago. I live among those born and bred to this place, descendants of those who arrived shortly after the Point Elliott Treaty of 1855 opened the area for white settlement. These are farmers whose forebears subdued the river and built farmhouses like the one I now call home, and who draw their livelihood from the rich soil. I am drawn more to the natural wonders of this place than to the economic potential of my 6 acres of prime bottomland. I look about me with eyes different from theirs and concern myself not with crop production but with the sustainability of salmon.

The Skagit River, third largest of the great West Coast rivers, has gone the way of most of our waterways. It has been harnessed, tamed, and replumbed from top to bottom. Five dams on her upper reaches, all fitted with hydroelectric power plants, supply electricity to the Seattle metropolitan area, 60 miles to the south. A vast network of dikes, levees, and drainage ditches make possible settled life and the variety of agricultural pursuits for which the Valley is famous. Famous, too, are Skagit salmon, both in variety and, until not too long ago, abundance. The watershed is home to all six Pacific salmon species and to core populations of Puget Sound chinook, recently listed as "threatened" under the Endangered Species Act. With more than one-third of these imperiled fish calling the Skagit home, this watershed is linchpin to the regional recovery effort.

This is a dark time for salmon. All over the region once seemingly inexhaustible runs, in decline for decades, are in a state of collapse. I imagine the Salmon People gathered in their villages beneath the salt water, engaged in a serious discussion about what they should do. They remember how, long ago, they decided that every year they would give the human beings a great gift, the gift of their flesh. They told the humans that they would allow themselves to be taken as food only on condition that they be treated with care and with respect. If they were not treated as honored guests, they would cease to come, and the memory of that time of abundance would be all that remained. I see the Salmon People talking for a long time before deciding that some of them would withhold their gift. Others would continue their annual runs for a while longer in the hope that the humans will change their ways.

Human beings, too, are engaged in much conversation about what should be done to keep the salmon coming. All agree that it is a daunting task to find cures for the many ailments that afflict salmon. Salmon range from headwaters to the deep ocean, their life cycle taking them through the entire range of human land uses and activities. Their habitat needs are difficult for the rivers of today to provide: natural meandering stream beds with deep pools and eddies, cool, clean water, sediment-free gravel beds, and quiet side-channels. Salmon present us with a complicated problem, and there is not a shared perception of urgency, let alone acceptable solutions. And because this is only one of many important issues, our institutions are not yet prepared to face up to the broader questions of environmental degradation that are raised by salmon decline. Still, I wait for the day when we collectively agree to mobilize a whole-hearted rescue mission, using all our resources and ingenuity and all the institutions of our society. In my mind's eye I see every last one of us rising up to protect what we say we value and what we call the soul of our region.

But the world we live in is far from perfect, full of ambiguities and contradictions. There is little that lines up neatly and it is futile to expect perfection from our systems or ourselves. Those of us who work on behalf of the natural world know that disappointment, disillusionment, and grief are constant companions, with us every time we act, every time we decide that the world can be a better place. Nevertheless, there is so much more to this world than that defined by human failing.

I arrived in the Skagit three years ago. To come here, I pulled up deep roots in the inner-city neighborhood of Seattle where I had lived for a decade and a half and where I had never been happier. I need to be closer to my work, I told people. The move surprised my colleagues and friends—and disoriented me for several months. But it was a studied venture, taken after much reflection and designed to bring me closer to the natural world I was feeling more and more distant from. That place could serve as spiritual center is something that I have always known intellectually. I have known too that for me to live a fully grounded life I need to be grounded in nature, in a place that is special to me. But I come from a people whose ways did not show me where that place would be. And so I have been free to wander, all the while longing to be an inhabitant, one who is rooted, connected and committed to the welfare of home.

In my new home I am surrounded by the elements and by all manner of living things, in deep, close interaction with the natural world. Season by season, I become more familiar with the wind patterns, the passage of the moon, the harrier hawks that hunt the meadow, the song of the coyotes, the lighting on the distant mountains. In the fall I listen for the familiar sound of the snow geese and trumpeter swans. I am teaching myself to pay attention to all the small details, to become attuned to the subtleties and particularities of this place. There is a certain spot where I stand, in the soft early morning light, in the fog, in the pounding rain or the ever-present wind, to look out on the landscape. Sometimes, when I am mindful, not wondering where the dogs are or agitated about what the busy day holds, the majesty of the scene draws me in and it touches my heart. I feel that I am an intrinsic part of all that is around me. It is the experience of interconnection, the abstraction that I pursue made real and personal.

I was recently on a panel with my friend Larry Campbell, talking about local partnerships to protect salmon. Larry is Swinomish and schooled in the old oral tradition. He speaks from the heart, unrehearsed, his words conveying what is on his mind and what the spirit in the room moves him to say. He starts by identifying himself with his people and his place. Names his grandparents and parents and tells us where they were from. His paternal line connects him to the Skagit River upstream while his maternal side places him on Skagit Bay. "I am just like the salmon, at home in both the salt and the fresh water," he says. He shows us on the map where he was born and raised and where he lives now. Smiles his sweet smile and lets us know that we are on his ancestral land. He speaks for a long time, in the sweeping circular indigenous manner, telling us about a way of life that is no more, about the grief and dislocation his people experience daily because of the decline of the salmon, one loss among many. "You can't separate the Indian from the salmon," he says. He tells us that the elders have encouraged him to reach out to people because the salmon needs help so urgently. He asks that we expand our families to embrace salmon and all the other animals. That we include the rivers and streams, the trees and the grasses, the roots and the berries. He talks about care and respect for the earth and its creatures. He reminds us of our responsibilities and our obligation, as humans. He closes by blessing us. His is a different voice, a story counter to the prevailing one, a truth-teller who does not deal in blame or judgment. In another time he would be a respected elder and a wise teacher. Today, he journeys into often hostile and unreceptive terrain to speak to those whose views, experience, and style could not be more different from his own. I watch him as he stands before us. His is a voice that has been systematically silenced, ignored, buried. Thought of little value. Yet here he is, determined and gentle, telling us what he thinks we need to know. I have heard his message before and am always moved, but this time it stirs me profoundly. I know that if I do not hold on tightly I will throw my head back, raise my arms and wail, releasing decades and eons of anguish into that windowless meeting room.

Later, when it's my turn, I tell the assembled group about the watershed council I am part of. The council, with a broad and diverse membership of thirty-six organizations, has developed a basinwide approach to repairing and protecting salmon habitat and is changing the way things are being done on the ground. I'm leery of the success we have had because I have seen other hopeful, vigorous efforts implode or flounder, so I spend time describing how we built what we hope will be a solid foundation. We chose to address only one part of a multipart problem—and our work alone will not ensure the future of salmon. We spent time educating one another on the complexities of our task and came to a collective understanding of how best to proceed. I describe the council's three-point program: a science-based strategy for restoration, the education of council members, and a celebration of place. And I talk about what I believe to be the true mission, that of laying the foundation for a different future, seeking not quick fixes but deeper understanding and new alternatives. I see us a hundred years from now slowly and respectfully applying our knowledge, shaping and fine-tuning with thoroughness and patience. Together, living in and making use of our watershed in a way that keeps it alive. Honoring the gift of life by giving back—and not taking too much.

"What about the farmers?" someone asks. A man I recognize as a neighbor says, "I think farmers are the salmon's best friend." And we talk about the challenge of maintaining the economic viability of agricultural production, the hard work and the uncertainty in the face of global markets and international competition, and wonder what it would take to better accommodate the needs of salmon. I say, as is my job to say, "We took so much and changed so much when we first came here. We will need to replace some of what has been lost and return some of that which should never have been taken." Other questions and discussion follow, genuine and serious inquiry into taken-for-granted ways of thinking and interaction. The farmer says he's not against cooperation but that it is hard to be confident that you're not going to get screwed, that it won't turn into something quite different from how it started out. We agree that where there is familiarity and trust, much is possible. Absent trust, agreements are reduced to legal documents or subject to nit-picky bureaucratic regulations. My neighbor, the farmer, shakes his head and says: "We've got to make a living, but we better do something before it's too late."

A woman asks me about the council's decision-making process and we talk about the importance of commonly held principles, clear, adhered-to procedures and how the council's process requires that there be common agreement among all members. The council chose to name its method "common agreement" instead of consensus. One member's viewpoint was that the term consensus is broadly overused and misused and not applicable to secular settings. He feels strongly that consensus is a spiritual process, not a worldly one. In his experience participants, gathered to make decisions on behalf of their community, come to their choices under divine auspices. His view of consensus is that the process brings with it the power of a shared experience, one of the spirit, where participants are moved from the "I" of individual self-interest to the "we" of common purpose, and so have a deeper sense of obligation to carry out that decision.

The same woman wonders if there can be such a thing as common purpose in a world as diverse as ours, with problems as complex as those we face. We agree that it is hard to see an adequately compelling vision of the future. I tell her what E. F. Schumacher said: "It takes a genius to make things simple." Sometimes what is important is that we are willing to act our way into clearer thinking. Intentional acts that reflect our values help us cut through the confusion and connect us to each other and to what is real. Barry Lopez suggests that, if in despair, we should step onto wounded ground and plant trees. Later, Larry and I exchange a few words outside. He touches my arm and smiles: Salmon's got you. I don't have to ask what he means. Salmon is coming for my heart and will teach me what I need to know.

What I am learning is that there are many dimensions to being grounded. As I claim my new home and let it claim me, I move onto a new path from spectator to participant. Long the facilitator, the third-party go-between, I embrace the substance and all the messiness and ambiguities that come with it. As I engage the world more intimately, my life becomes more sensory, more tactile. I am mindful of the details. I pay attention to sounds and smells, textures and shapes and colors. Outwardly, my house is messier than it ever was, my clothes are no longer city-crisp, and the dirt of my kitchen garden is often under my nails. Inwardly, there is less despair, more joy. I am beginning to uncover my heart and find the sun.

In talking about the barriers to women's equality, Gloria Steinem likens the female spirit—relational, intuitive, nurturing, caring, and cooperative— to a garden of sun plants that has been grown in the shade. They survive but at the cost of their true form. As a child in apartheid-era South Africa, my world was shadowed by patriarchy, dominance, and violence. I saw the high price to be paid for being different and found a way to leave, cutting myself off from heritage, family, and birthplace. As a woman professional trying to get along in an intellectualized, impersonal, man-dominated work world, I took on what felt like a genderless persona, one that gained me access and provided me safe passage but stifled intuition and caring.

Now, past the midpoint of my life, I no longer live in the shade. I see myself in a circle of resilient, determined, loving women, the grandmothers, the mothers, and the aunties who tell the new stories of community, the stories that reconnect us to each other, show us what right living is about, stories that bind us closer and closer to our place as the people of the Skagit. I hear the music and the happy laughter of the dancers as they celebrate the changing of the seasons, welcome in the harvest, and honor the first coming of the salmon. And I smell the flesh of that salmon cooking over the cedar logs.

In nature, I become myself. Through salmon, I may yet come to be at home.

REFERENCES

Bateson, Mary, *Composing a Life.* New York: Penguin, 1989.

Drengson, Alan, and Uichi Inoue, eds., *The Deep Ecology Movement: An Introductory Anthology.* Berkeley, California: North Atlantic Books, 1995.

Durning, Alan Thein, *This Place on Earth: Home and the Practice of Permanence.* Seattle, Washington: Sasquatch Books, 1996.

Heifetz, Ronald A., *Leadership Without Easy Answers.* Cambridge, Massachusetts: Belknap Press, 1994.

House, Freeman, *Totem Salmon: Life Lessons from Another Species.* Boston: Beacon Press, 1999.

Appendix: Selected Resources in Collaborative Conservation

Alex Conley

Many voices have played a part in the development of what this book refers to as collaborative conservation, and what others refer to as partnerships, consensus groups, community-based collaboratives, watershed councils, community-based ecosystem management, grassroots ecosystem management, and coordinated resources management. The first section of this bibliography introduces some of the principal lines of thought that have contributed to the development of collaborative conservation, and lists sources where the interested reader can learn more about each area. The second section lists sources that specifically address recent developments in collaborative conservation in the United States.

This bibliography was initially developed for the Consortium for Research and Assessment of Community-Based Collaboratives, with the support of The University of Arizona's Udall Center for Studies in Public Policy. Thanks are due to the many people whose suggestions have been incorporated into this work.

Those interested in a more comprehensive list of sources with more detailed descriptions of each subject area may want to acquire the expanded version from the Udall Center. Mail requests to Publications Editor, Udall Center for Studies in Public Policy, 803 E. First St., Tucson, AZ 85719, or e-mail to udallctr@u.arizona.edu.

I. Sources Informing Collaborative Conservation

International Experience with Community-Based Conservation

Many involved in collaborative conservation in the United States refer to it as an idea that originated overseas and is now being brought home. Both "participatory development" and "community-based conservation" are concepts that are widely used in the international development arena.

Ascher, William. *Communities and Sustainable Forestry in Developing Countries.* San Francisco: ICS Press, 1995.

Baland, Jean Marie, and Jean Philippe Platteau. *Halting Degradation of Natural Resources: Is There a Role for Rural Communities?* New York: Oxford University Press, 1996.

Chambers, Robert. *Whose Reality Counts? Putting the First Last.* London: Intermediate Technology, 1997.

Food and Agriculture Organization. Community Forestry Web Site. <www.fao.org>

Getz, Wayne M., Louise Fortmann, David Cumming, Johan du Toit, Jodi Hilty, Rowan Martin, Michael Murphree, Norman Owen-Smith, Anthony M. Starfield, and Michael I. Westphal. "Sustaining Natural and Human Capital: Villagers and Scientists." *Science* 283 (1999): 1855–1856.

Liz Claiborne and Art Ortenberg Foundation. *The View from Airlie: Community-Based Conservation in Perspective.* New York: Liz Claiborne and Art Ortenberg Foundation, 1993.

Western, David, and R. Michael Wright, eds. *Natural Connections: Perspectives in Community-Based Conservation.* Washington, D.C.: Island Press, 1994.

Common Property Management and Comanagement

Recent research into common property management systems emphasizes the often-effective role local institutions have played in the sustainable management of natural resources in virtually all parts of the world. This research has led many to reassess the way that "the tragedy of the commons" has been used to justify state control of natural resources, and is often used to give credence to assertions that community involvement can improve the management of natural resources. Comanagement—the sharing of decision-making authority by local resource users and state and national governments—is now being broadly promoted throughout the world.

Baden, John A., and Douglas S. Noonan, eds. *Managing the Commons.* Bloomington: Indiana University Press, 1998.

Berkes, Fikret, ed. *Common Property Resources: Ecology and Community-Based Sustainable Development.* New York: Belhaven Press, 1989.

Bromley, Daniel W., and David Feeny, eds. *Making the Commons Work: Theory, Practice, and Policy.* San Francisco: ICS Press, 1992.

International Association for the Study of Common Property Web Site. <www.indiana.edu/~iascp>

Ostrom, Elinor. *Governing the Commons: The Evolution of Institutions for Collective Action.* New York: Cambridge University Press, 1990.

Pinkerton, Evelyn, ed. *Co-operative Management of Local Fisheries: New Directions for Improved Management and Community Development.* Vancouver: University of British Columbia Press, 1989.

Democratic Theory, Procedural Justice, and Social Capital

Many democratic theorists promote the idea of participatory democracy, and collaborative conservation efforts are frequently used as examples of this new form of governance, which is built on Jeffersonian ideals. Procedural justice is the idea that people who participate in rule making are more likely to accept unfavorable outcomes based on those rules. Social capital—the "features of social organization such as networks, norms, and social trust that facilitate coordination and cooperation for mutual benefit" (Putnam 1995)—is often identified as both a prerequisite for and a product of collaborative efforts.

Barber, Benjamin. *Strong Democracy.* Berkeley and Los Angeles: University of California Press, 1984.

Dasgupta, Partha, and Ismail Serageldin, eds. *Social Capital: A Multifaceted Perspective.* Washington, D.C.: World Bank, 1999.

DeWitt, John. *Civic Environmentalism: Alternatives to Regulation in States and Communities.* Washington, D.C.: CQ Press, 1994.

Dryzek, John S. *Discursive Democracy: Politics, Policy, and Political Science.* New York: Cambridge University Press, 1990.

Lawrence, Rick L., Steven E. Daniels, and George H. Stankey. "Procedural Justice and Public Involvement in Natural Resources Decision Making." *Society and Natural Resources* 10, no. 6 (1997): 577–589.

Mathews, Forrest David. *Politics for People: Finding a Responsible Public Voice.* Urbana: University of Illinois Press, 1994.

Mathews, Freya, ed. *Ecology and Democracy.* Portland, Oregon: Frank Cass, 1996.

Morone, James A. *The Democratic Wish: Popular Participation and the Limits of American Government.* New York: Basic Books, 1990.

Pateman, Carole. *Participation and Democratic Theory.* Cambridge, England: University Press, 1970.

Press, Daniel. *Democratic Dilemmas in the Age of Ecology.* Durham, N.C.: Duke University Press, 1994.

Putnam, Robert D. "Bowling Alone: America's Declining Social Capital." *Journal of Democracy* 6, no. 1 (1995): 65–78.

Rescher, Nicholas. *Pluralism: Against the Demand for Consensus.* New York: Oxford University Press, 1993.

Tyler, Tom R. "The Psychology of Procedural Justice: A Test of the Group-Value Model." *Journal of Personality and Social Psychology* 57, no. 5 (1989): 830–838.

Weber, Edward P. "The Question of Accountability in Historical Perspective: From Jackson to Contemporary Grassroots Ecosystem Management." *Administration and Society* 31, no. 4 (1999): 451–494.

———. *Pluralism by the Rules: Conflict and Cooperation in Environmental Regulation.* Washington, D.C.: Georgetown University Press, 1998.

Williams, Bruce Alan, and Albert R. Matheny. *Democracy, Dialogue, and Environmental Disputes.* New Haven, Connecticut: Yale University Press, 1995.

Existing Public Participation Mechanisms

This body of literature looks more closely at how existing mechanisms of public participation have functioned in environmental and natural resources planning efforts.

Behan, R. W. "A Plea for Constituency-Based Management." *American Forests* 97 (1988): 46–48.

Blahna, Dale J., and Susan Yonts-Shepard. "Public Involvement in Resource Planning: Towards Bridging the Gap between Policy and Implementation." *Society and Natural Resources* 2, no. 3 (1989): 209–227.

Cortner, Hanna J., and Margaret Shannon. "Embedding Public Participation in Its Political Context." *Journal of Forestry* 91, no. 7 (1993): 14–16.

Gericke, Kevin L., and Jay Sullivan. "Public Participation and Appeals of Forest Service Plans: An Empirical Examination." *Society and Natural Resources* 7, no. 2 (1994): 125–135.

Knopp, Timothy B., and Elaine S. Caldbeck. "The Role of Participatory Democracy in Forest Management." *Journal of Forestry* 88, no. 5 (1990): 13–18.

McMullin, Steve L., and Larry A. Nielsen. "Resolution of Natural Resources Allocation Conflicts Through Effective Public Involvement." *Policy Studies Journal* 19 (1991): 553–559.

Mohai, Paul. "Public Participation and Natural Resources Decision Making: The Case of the RARE II Decisions." *Natural Resources Journal* 27, no. 1 (1987): 123–155.

Richard, Tim, and Sam Burns. "Beyond 'Scoping': Citizens and San Juan National Forest Managers, Learning Together." *Journal of Forestry* 96, no. 4 (1998): 39–43.

Sample, V. Alaric. "A Framework for Public Participation in Natural Resource Decisionmaking." *Journal of Forestry* 91, no. 7 (1993): 22–27.

Shands, William E. 1991. "Reaching Consensus on National Forest Use." *Forum for Applied Research and Public Policy* 6, no. 3 (1991): 18–23.

Shannon, Margaret. "Building Trust: The Formation of a Social Contract." In *Community and Forestry: Continuities in the Sociology of Natural Resources,* edited by Robert G. Lee, Donald R. Field, and William R. Burch Jr. Boulder, Colorado: Westview Press, 1990.

Sirmon, Jeff, William E. Shands, and Chris Liggett. "Communities of Interests and Open Decisionmaking." *Journal of Forestry* 91, no. 7 (1993): 17–21.

Wellman, J. Douglas, and Terence J. Tipple. "Public Forestry and Direct Democracy." *The Environmental Professional* 12, no. 1 (1990): 77–86.

Theories of Collaboration

Collaborative conservation draws on theories of collaboration that have been developed both in the fields of organizational behavior, public administration, and community psychology and through practical experiences with collaborative processes in business, government, and nonprofit sectors.

Chrislip, David D., and Carl E. Larson. *Collaborative Leadership: How Citizens and Civic Leaders Can Make a Difference.* San Francisco: Jossey-Bass, 1994.

Gray, Barbara. *Collaborating: Finding Common Ground for Multiparty Problems.* San Francisco: Jossey-Bass, 1989.

Huxam, Chris. *Creating Collaborative Advantage.* London: Sage, 1996.

London, Scott. *Collaboration and Community.* Pew Partnership for Civic Change. Web Site. <www.west.net/~insight/london/ppcc.htm>

Mattessich, Paul W., and Barbara R. Monsey. *Collaboration: What Makes It Work—A Review of Research Literature on Factors Influencing Successful Collaboration.* St. Paul, Minnesota: Amherst H. Wilder Foundation, 1992.

McCann, Joseph E. "Design Guidelines for Social Problem-Solving Interventions." *Journal of Applied Behavioral Sciences* 19, no. 2 (1983): 177–192.

Waddock, S. A. "Understanding Social Partnerships: An Evolutionary Model of Partnership Organizations." *Administration and Society* 21, no. 1 (1989): 78–100.

Winer, M., and K. Ray. *Collaboration Handbook: Creating, Sustaining and Enjoying the Journey.* St. Paul, Minnesota: Amherst H. Wilder Foundation, 1996.

Community Development

In recent decades, the field of community development has increasingly focused on how to foster the basic conditions that make for successful communities. Visioning, strategic planning, and other tools now being applied to collaborative conservation have been widely used in efforts to increase social capital, build community capacity, and improve the quality of life in communities of all sizes. On a more theoretical level, rural sociologists have helped to redefine how community well-being is assessed and to increase understanding of the dynamics of poverty, exploitation, and internal colonialism that many collaborative efforts strive to redress.

Community Development Society Web Site. <www.comm-dev.org>

Ford Foundation. Exploring Conservation-Based Development Web Site. <www.explorecbd.org>

Frentz, Irene, Sam Burns, Donald E. Voth, and Charles Sperry. *Rural Development and Community-Based Forest Planning and Management: A New, Collaborative Paradigm.* USDA National Research Institute Project 96-35401-3393. Fayetteville: University of Arkansas, 1999.

Howe, Jim, Edward McMahon, and Luther Propst. *Balancing Nature and Commerce in Gateway Communities.* Washington, D.C.: Island Press, 1997.

Kingsley, G. T., J. B. McNeely, and J. O. Gibson. *Community Building: Coming of Age.* Washington, D.C.: The Urban Institute, 1996.

Kusel, Jonathan. "Well-Being in Forest Dependant Communities, Part I: A New Approach." In *Sierra Nevada Ecosystem Project: Final Report to Congress, Vol. II.* Davis: University of California, Centers for Water and Wildland Resources, 1996.

Lee, Robert G., Donald R. Field, and William R. Burch Jr., eds. *Community and Forestry: Continuities in the Sociology of Natural Resources.* Boulder, Colorado: Westview Press, 1990.

Moore, Carl, Gianni Longo, and Patsy Palmer. "Visioning." In *The Consensus Building Handbook: A Comprehensive Guide to Reaching Agreement,* edited by Lawrence Susskind, Sarah McKearnan, and Jennifer Thomas-Larmer. Thousand Oaks, California: Sage Publications, 1999.

Peluso, Nancy Lee, Craig R. Humphrey, and Louise P. Fortmann. "The Rock, the Beach and the Tide Pool: People and Poverty in Natural Resource–Dependent Areas." *Society and Natural Resources* 7, no. 1 (1994): 23–28.

Potapchuck, W. R., and C. G. Polk. *Building the Collaborative Community.* Washington, D.C.: Program for Community Problem Solving, National Civic League, 1994.

Sonoran Institute Web Site. <www.sonoran.org>

Discussions of Place and Community

A number of scholars have looked at the role of "community" in shaping our sense of social responsibility and interdependence and the way "sense of place" informs our relationship to the landscapes in which we live. Their work has been broadly influenced by theories of democracy and social capital, literary ideas about how our sense of community and place shape us, and populist interest in neighborliness and small-town self-governance. Collaborative conservation is often seen as a natural extension of this community-based vision.

Basso, Keith H. *Wisdom Sits in Places: Landscape and Language among the Western Apache.* Albuquerque: University of New Mexico Press, 1996.

Brandenburg, Andrea M., and Matthew S. Carroll. "Your Place or Mine? The Effect of Place Creation on Environmental Values and Landscape Meanings." *Society and Natural Resources* 8, no. 5 (1995): 381–398.

Etzioni, Amitai. *An Immodest Agenda: Rebuilding America before the 21st Century.* New York: New Press, 1983.

Feld, Steven, and Keith H. Basso, eds. *Senses of Place.* Santa Fe: School of American Research Press, 1996.

Kemmis, Daniel. *Community and the Politics of Place.* Norman: University of Oklahoma Press, 1990.

Sagoff, Mark. *The Economy of the Earth: Philosophy, Law, and the Environment.* Cambridge: Cambridge University Press, 1988.

Devolution and Market-Based Approaches

Federal agencies are often portrayed as inefficient bureaucracies, and many authors promote devolving federal powers to more local levels and/or using alternate management strategies—many of them based in free market approaches. Many of these arguments are used to justify the use of alternative, collaborative approaches.

Foundation for Research on Economics and the Environment Web Site. <www. free-eco.org>

Fretwell, Holly Lippke. *Public Lands Forests: Do We Get What We Pay For?* Bozeman, Montana: Political Economy Research Center, 1999.

Nelson, Robert H. *Public Lands and Private Rights: The Failure of Scientific Management.* Lanham, Maryland: Rowman & Littlefield, 1995.

O'Toole, Randal. *Reforming the Forest Service.* Washington, D.C.: Island Press, 1988.

Political Economy Research Center Web Site. <www.perc.org>

The Sagebrush Rebellion and the Wise-Use Movement

Collaborative efforts often exist in political climates where there is a sometimes uneasy coexistence of environmental values and those of so-called sagebrush rebels and wise-use advocates. Understanding wise use helps make sense of the political scene in which collaborative conservation might thrive.

Arnold, Ron. *Ecology Wars: Environmentalism as if People Mattered.* 1st ed. Bellevue, Washington: Free Enterprise Press, 1987.

Brick, Philip D., and R. McGreggor Cawley, eds. *A Wolf in the Garden: The Land Rights Movement and the New Environmental Debate.* Lanham, Maryland: Rowman & Littlefield, 1996.

Cawley, R. McGreggor. *Federal Land, Western Anger: The Sagebrush Rebellion and Environmental Politics.* Lawrence: University Press of Kansas, 1993.

Dagget, Dan. "Getting Out of the Cow Business: Nevada Sagebrush Rebels Shift Gears." *Chronicle of Community* 1, no. 2 (1997): 5–15.

Echeverria, John D., and Raymond Booth Eby, eds. *Let the People Judge: Wise Use and the Private Property Rights Movement.* Washington, D.C.: Island Press, 1995.

Helvarg, David. *The War Against the Greens: The "Wise-Use" Movement, the New Right, and Anti-Environmental Violence.* San Francisco: Sierra Club Books, 1994.

McCarthy, James. "Environmentalism, Wise Use and the Nature of Accumulation in the Rural West." In *Remaking Reality: Nature at the Millennium,* edited by Bruce Braun and Noel Castree. New York: Routledge, 1998.

Switzer, Jacqueline Vaughn. *Green Backlash: The History and Politics of Environmental Opposition in the U.S.* Boulder, Colorado: Lynne Rienner Publishers, 1997.

Alternative Dispute Resolution

Alternative dispute resolution (ADR) has its roots in international peacemaking and labor negotiations but is now commonly used in efforts to resolve environmental and natural resource policy disputes.

Amy, Douglas J. *The Politics of Environmental Mediation.* New York: Columbia University Press, 1987.

Bingham, Gail. *Resolving Environmental Disputes: A Decade of Experience*. Washington, D.C.: Conservation Foundation, 1986.

Blackburn, J. Walton, and Willa M. Bruce, eds. *Mediating Environmental Conflicts: Theory and Practice*. Westport, Connecticut: Quorum Books, 1995.

Buckle, Leonard G., and Suzann R. Thomas-Buckle. "Placing Environmental Mediation in Context: Lessons from 'Failed' Mediations." *Environmental Impact Assessment Review* 6, no. 1 (1986): 55–70.

Burgess, Heidi, and Guy Burgess. "Constructive Confrontation: A Transformative Approach to Intractable Conflicts." *Mediation Quarterly* 13, no. 4 (1996): 305–322.

Carpenter, Susan. "Dealing with Environmental and Other Public Disputes." In *Community Mediation: A Handbook for Practitioners and Researchers*, edited by Karen Grover Duffy, James W. Grosch, and Paul V. Olczak. New York: The Guilford Press, 1991.

Crowfoot, James E., and Julia Marie Wondolleck. *Environmental Disputes: Community Involvement in Conflict Resolution*. Washington, D.C.: Island Press, 1990.

Ellickson, Robert C. *Order Without Law: How Neighbors Settle Disputes*. Cambridge, Massachusetts: Harvard University Press, 1991.

Emerson, Kirk, Richard Yarde, and Tanya Heikkila, eds. *Environmental Conflict Resolution in the West: Conference Proceedings*. Tucson: The Udall Center for Studies in Public Policy, University of Arizona, 1997.

Fisher, Roger, and William Ury. *Getting to Yes: Negotiating Agreement Without Giving In*. Boston: Houghton Mifflin, 1981.

Modavi, Neghin. "Mediation of Environmental Conflicts in Hawaii: Win-Win or Cooptation?" *Sociological Perspectives* 39, no. 2 (1996): 301–316.

Ozawa, Connie P. *Recasting Science: Consensual Procedures in Public Policy Making*. Boulder, Colorado: Westview Press, 1991.

Sipe, Neil G. "An Empirical Analysis of Environmental Mediation." *Journal of the American Planning Association* 64, no. 3 (1998): 275–285.

Susskind, Lawrence, and Jeffrey L. Cruikshank. *Breaking the Impasse: Consensual Approaches to Resolving Public Disputes*. New York: Basic Books, 1987.

Susskind, Lawrence, Sarah McKearnen, and Jennifer Thomas-Larmer, eds. *The Consensus Building Handbook: A Comprehensive Guide to Reaching Agreement*. Thousand Oaks, California: Sage Publications, 1999.

Wondolleck, Julia Marie. *Public Lands Conflict and Resolution: Managing National Forest Disputes*. New York: Plenum Press, 1988.

Wondolleck, Julia M., Nancy J. Manring, and James E. Crowfoot. "Teetering at the Top of the Ladder: The Experience of Citizen Group Participants in Alternative Dispute Resolution Processes." *Sociological Perspectives* 39, no. 2 (1996): 249–262.

Ecosystem and Watershed Management

Collaborative conservation also relies heavily on ecosystem, watershed, and adaptive management approaches. The first two emphasize the need to coordinate decision making across different land ownerships and administrative boundaries, and often promote collaborative approaches as a way to achieve this goal. Adaptive management emphasizes an experimental, iterative approach to decision making.

Cortner, Hanna J., and Margaret A. Moote. *The Politics of Ecosystem Management.* Washington, D.C.: Island Press, 1999.

Griffin, C. B. "Watershed Councils: An Emerging Form of Public Participation in Natural Resource Management." *Journal of the American Water Resources Association* 35, no. 3 (1999): 505–518.

Grumbine, R. Edward. "What Is Ecosystem Management?" *Conservation Biology* 8, no. 1 (1994): 27–38.

Gunderson, Lance H., C. S. Holling, and Stephen S. Light, eds. *Barriers and Bridges to the Renewal of Ecosystems and Institutions.* New York: Columbia University Press, 1995.

Kenney, Douglas S. "Historical and Sociopolitical Context of the Western Watersheds Movement." *Journal of the American Water Resources Association* 35, no. 3 (1999): 493–503.

Kenney, Douglas, et al. *The New Watershed Source Book.* Boulder, Colorado: Natural Resources Law Center, University of Colorado School of Law, 2000.

Knight, Richard L., and Peter B. Landres, eds. *Stewardship Across Boundaries.* Washington, D.C.: Island Press, 1998.

Kusel, Jonathan, Sam C. Doak, Susan Carpenter, and Victoria E. Sturtevant. "The Role of the Public in Adaptive Ecosystem Management." In *Sierra Nevada Ecosystem Project: Final Report to Congress,* Vol. II. Davis: University of California, Centers for Water and Wildland Resources, 1996.

Lee, Kai N. *Compass and Gyroscope: Integrating Science and Politics for the Environment.* Washington, D.C.: Island Press, 1993.

McLain, Rebecca J., and Robert G. Lee. "Adaptive Management: Promises and Pitfalls." *Environmental Management* 20, no. 4 (1996): 437–448.

Yaffee, Steven L., Ali F. Phillips, Irene C. Frentz, Paul W. Hardy, Sussanne M. Maleki, and Barbara E. Thorpe. *Ecosystem Management in the United States.* Washington, D.C.: Island Press, 1996.

II. Sources Discussing Collaborative Conservation

Overviews of Collaborative Conservation in the United States

Overviews of the use and development of collaborative conservation can be found in everything from political speeches and newspapers to academic publications.

American Forests. "Local Voices, National Issues." *American Forests* 103, no. 4 (1998).

Bernard, Ted, and Jora Young. *The Ecology of Hope: Communities Collaborate for Sustainability.* Gabriola Island, B.C.: New Society Publishers, 1997.

Brendler, Thomas, and Henry Carey. "Community Forestry, Defined." *Journal of Forestry* 96, no. 3 (1998): 21–23.

Carr, Deborah S., Steven W. Selin, and Michael A. Schuett. "Managing Public Forests: Understanding the Role of Collaborative Planning." *Environmental Management* 22, no. 5 (1998): 767–776.

Cestero, Barb. *Beyond the Hundredth Meeting: A Field Guide to Collaborative Conservation on the West's Public Lands.* Tucson, Arizona: The Sonoran Institute, 1999.

Consultative Group on Biological Diversity. *A Report from Troutdale: Community-Based Strategies in Forest Stewardship and Sustainable Economic Development.* San Francisco: Consultative Group on Biological Diversity, 1998.

Coughlin, Christine W., Merrick L. Hoben, Dirk W. Manskopf, and Shannon W. Quesada. "A Systematic Assessment of Collaborative Resource Management Partnerships." Master's Project, University of Michigan, 1999.

Dagget, Dan. *Beyond the Rangeland Conflict: Toward a West That Works.* Layton, Utah: Gibbs-Smith, 1995.

Daniels, Steven E., and Gregg B. Walker. "Collaborative Learning: Improving Public Deliberation in Ecosystem-Based Management." *Environmental Impact Assessment Review* 16 (1996): 71–102.

Daniels, Steven E., and Gregg B. Walker. *Working through Environmental Policy Conflicts: The Collaborative Learning Approach.* New York: Praeger, 2000.

Fairfax, Sally, Lynn Huntsinger, and Carmel Adelburg. "Lessons from the Past: Old Conservation Models Provide New Insight into Community-Based Land Management." *Forum for Applied Research and Public Policy* 14, no. 2 (1999): 84–88.

Gray, G. J., Maia J. Enzer, and Jonathan Kusel, eds. *Understanding Community-Based Ecosystem Management in the United States.* New York: Haworth Press, 2001.

High Country News. *Index to High Country News Stories on Consensus Groups.* <www.hcn.org>

Jones, Lisa. "Howdy neighbor! As a Last Resort, Westerners Start Talking to Each Other." *High Country News* 28, no. 9 (1996): 1, 6–8.

Kusel, J., G. J. Gray, and M. J. Enzer, eds. *Proceedings of the Lead Partnership Group, Northern California/Southern Oregon Roundtable on Communities of Place, Partnerships, and Forest Health.* Washington, D.C.: American Forests/Forest Community Research, 1996.

Krueger, William C. "Building Consensus for Rangeland Uses." *Rangelands* 14, no. 1 (1992): 38–40.

McKinney, Matthew. "Governing Western Resources: A Confluence of Ideas." *Rendezvous: The Humanities in Montana* 2, no. 2 (1999): 4–11.

Selin, Steve, and Deborah Chavez. "Developing a Collaborative Model for Environmental Planning and Management." *Environmental Management* 19, no. 2 (1995): 189–195.

How-To Guides

As interest in collaborative approaches grows, more and more handbooks and guides are written to assist people facilitating or participating in collaborative processes.

Anderson, E. William, and Robert C. Baum. "How to Do Coordinated Resource Management Planning." *Journal of Soil and Water Conservation* 43, no. 3 (1988): 216–220.

Clark, Jo. *Watershed Partnerships: A Strategic Guide for Local Conservation Efforts in the West.* Denver: Western Governors' Association, 1997.

Cleary, C. Rex, and Dennis Phillippi. *Coordinated Resource Management: Guidelines for All Who Participate.* Denver: Society for Range Management, 1993.

Environmental Protection Agency. *Community-Based Environmental Protection: A Resource Book for Protecting Ecosystems and Communities.* EPA 230-B-96-003. Washington, D.C.: Environmental Protection Agency, 1997.

Luscher, Kathy. *Starting Up: A Handbook for New River and Watershed Organizations.* Portland, Oregon: River Network, 1996.

Montana Consensus Council. *Resolving Public Disputes: A Handbook on Building Consensus.* Helena, Montana: Montana Consensus Council, 1998.

Moote, Margaret A. *The Partnership Handbook.* <www.ag.arizona.edu/partners>, 1996.

Oregon State University Extension Service. *Watershed Stewardship: A Learning Guide.* Corvallis: Oregon State University Extension Service, 1998.

Paulson, Deborah D., and Katherine M. Chamberlin. *Guidelines and Issues to Consider in Planning a Collaborative Process.* Laramie: Institute for Environment and Natural Resources, University of Wyoming, 1998.

Tarnow, K., P. Watt, and D. Silverberg. *Collaborative Approaches to Decision Making and Conflict Resolution for Natural Resource and Land Use Issues: A Handbook for Land Use Planners, Resource Managers and Resource Management Councils.* Salem: Oregon Department of Land Conservation and Development, 1996.

Case Studies

Case studies provide an excellent way of understanding how collaborative conservation is developing on the ground, and many more are being written each year.

Braxton Little, Jane. "The Feather River Alliance: Restoring Creeks and Communities in the Sierra Nevada." *Chronicle of Community* 2, no. 1 (1997): 5–14.

———. "The Whiskey Creek Group: Where Consensus Is Not a Goal and the Forest Service Is Not the Devil." *Chronicle of Community* 3, no. 3 (1999): 5–11.

Callister, Deborah Cox. *Community & Wild Lands Futures: A Pilot Project in Emery County, Utah.* Salt Lake City: Coalition for Utah's Future Project 2000, 1995.

Cestero, Barb. "From Conflict to Consensus? A Social and Political History of Environmental Collaboration in the Swan Valley, Montana." Masters Thesis, University of Montana, 1997.

Chasteen, Bonnie. "Conscience, With a Price Tag: Eco-labels and Niche Brands Help Proven Stewards Stay on the Land." *Chronicle of Community* 3, no. 2 (1999): 15–24.

Chisholm, Graham. "Tough Towns: The Challenge of Community-Based Conservation." In *A Wolf in the Garden: The Land Rights Movement and the New Environmental Debate*, edited by Philip D. Brick and R. McGreggor Cawley. Lanham, Maryland: Rowman and Littlefield Publishers, 1996.

Covert, John, and Sarah Van de Wetering. "Saving the Ranch, Saving the Land: Ranchers and Conservation Buyers Seek Alternatives to Subdividing." *Chronicle of Community* 2, no. 3 (1998): 17–25.

Daniels, Steven E., Gregg B. Walker, Matthew S. Carroll, and Keith A. Blatner. "Using Collaborative Learning in Fire Recovery Planning." *Journal of Forestry* 94, no. 8 (1996): 4–9.

Duane, Timothy P. "Community Participation in Ecosystem Management." *Ecology Law Quarterly* 24, no. 4 (1997): 771–797.

Hasselstrom, Linda. "Rising from the Condos: Community Land Trust and Longtime Residents Team Up to Ensure Affordable Housing in Jackson, Wyoming." *Chronicle of Community* 2, no. 3 (1998): 5–16.

House, Freeman. *Totem Salmon: Life Lessons from Another Species.* Boston: Beacon Press, 1999.

KenCairn, Brett. "Peril on Common Ground: The Applegate Experiment." In *A Wolf in the Garden: The Land Rights Movement and the New Environmental Debate,* edited by Philip D. Brick and R. McGreggor Cawley. Lanham, Maryland: Rowman and Littlefield Publishers, 1996.

Krist, John. "Seeking Common Ground: Water Lubricates Armistice among Traditional Foes in California." *Chronicle of Community* 3, no. 3 (1999): 12–23.

Moote, Margaret A., Mitchel P. McClaran, and Donna K. Chickering. "Theory in Practice: Applying Participatory Democracy Theory to Public Land Planning." *Environmental Management* 21, no. 6 (1997): 877–889.

Richard, Tim, and Sam Burns. *Ponderosa Pine Forest Partnership: Forging New Relationships to Restore a Forest.* Durango, Colorado: Fort Lewis College Office of Community Services, 1998.

Shelly, Steve. "Making a Difference on the Ground: Colorado's Ponderosa Pine Partnership Shows How It Can Be Done." *Chronicle of Community* 3, no. 1 (1998): 37–39.

Smith, Melinda. "The Catron County Citizens' Group: A Case Study in Community Collaboration." In *The Consensus Building Handbook: A Comprehensive Guide to Reaching Agreement,* edited by Lawrence Susskind, Sarah McKearnan, and Jennifer Thomas-Larmer. Thousand Oaks, California: Sage Publications, 1999.

Snow, Donald. "Strong Foundations: Can the Beacons Project Save Montana's Small Towns?" *Chronicle of Community* 1, no. 3 (1997): 13–23.

Sturtevant, V. E., and J. I. Lange. *Applegate Partnership Case Study: Group Dynamics and Community Context.* Ashland: Southern Oregon State College (for the U.S. Forest Service Pacific Northwest Research Station), 1995.

Van de Wetering, Sarah. "Doing It the Moab Way: A Public Land Partnership at Sand Flats (UT)." *Chronicle of Community* 1, no. 1 (1996): 5–16.

———. "A Seamless Canyon: Zion National Park and Springdale, Utah, Discover the Powers of Partnership." *Chronicle of Community* 3, no. 2 (1999): 5–14.

———. "'Enlightened Self-Interest': Wyoming Experiments with Coordinated Resource Management." *Chronicle of Community* 1, no. 2 (1997): 17–25.

Wolf, Tom. "Bienvenidos a San Luis: A Colorado Town Melds Faith with Community Activism, but Its Goals Remain Elusive." *Chronicle of Community* 2, no. 1 (1997): 15–25.

Criticisms

Many criticisms of collaborative conservation are now being raised. Most come from environmental activists who perceive collaborative efforts as inefficient and/or as dangerous attempts to assert local—often industry—control over natural resources.

Blumberg, Louis, and Darrell Knuffke. "Count Us Out: Why the Wilderness Society Opposed the Quincy Library Group Legislation." *Chronicle of Community* 2, no. 2 (1998): 41–44.

Britell, Jim. *Essays #10–12.* <www.britell.com>

Coggins, George C. "Regulating Federal Natural Resources: A Summary Case against Devolved Collaboration." *Ecology Law Quarterly* 25, no. 4 (1998): 602–610.

Coglianese, Cary. "The Limits of Consensus." *Environment* 41, no. 3 (1999): 28–33.

Kenney, Douglas S. *Arguing About Consensus: Examining the Case Against Western Watershed Initiatives and Other Collaborative Groups Active in Natural Resources Management.* Boulder: Natural Resources Law Center, University of Colorado School of Law, 2000.

Leach, Melissa, Robin Mearns, and Ian Scoones. "Challenges to Community-Based Sustainable Development." *Institute of Developmental Studies Bulletin* 28, no. 4 (1997): 4–14.

McCloskey, Michael. "Local Communities and the Management of Public Forests." *Ecology Law Quarterly* 25, no. 4 (1999): 624–629.

———. "The Skeptic: Collaboration Has Its Limits." High Country News 28, no. 9 (1996): 7.

McLain, Rebecca J., and Eric Jones. *Challenging "Community" Definitions in Sustainable Natural Resources Management: The Case of Wild Mushroom Harvesting in the USA.* Gatekeeper Series no. 68. London: International Institute for Environment and Development, 1997.

Southern Utah Wilderness Association. "Why One Advocacy Group Steers Clear of Consensus Efforts." *High Country News* 26, no. 10 (1994).

Evaluating Collaborative Conservation

Interest in evaluating collaborative conservation—both in terms of assessing its effectiveness and in terms of identifying facilitating and inhibiting factors—is growing.

Blumberg, Louis. "Preserving the Public Trust." *Forum for Applied Research and Public Policy* 14, no. 2 (1999): 89–93.

Brendler, Thomas, and Shirl Crosman. *The Federal Advisory Committee Act: Implications for Public Involvement on the National Forests.* Santa Fe: The Forest Trust, 1995.

Kagan, Robert A. "Political and Legal Obstacles to Collaborative Ecosystem Planning." *Ecology Law Quarterly* 24, no. 4 (1997): 871–875.

Kenney, Douglas S., and William B. Lord. *Analysis of Institutional Innovation in the*

Natural Resources and Environmental Realm. Boulder: Natural Resources Law Center, University of Colorado School of Law, 1999.

Lynch, Sheila. "The Federal Advisory Committee Act: An Obstacle to Ecosystem Management by Federal Agencies?" *Washington Law Review* 71 (1996): 431–459.

Moote, M. A., and M. P. McClaran. "Viewpoint: Implications of Participatory Democracy for Public Land Planning." *Journal of Range Management* 50, no. 5 (1997): 473–481.

Nickelsburg, Stephen M. "Mere Volunteers? The Promise and Limits of Community-Based Environmental Protection." *Virginia Law Review* 84 (1998): 1371–1409.

Paulson, Deborah D. "Collaborative Management of Public Rangeland in Wyoming: Lessons in Co-management." *Professional Geographer* 50, no. 3 (1998): 301–315.

Williams, Ellen M., and Paul V. Ellefson. "Going into Partnership to Manage a Landscape." *Journal of Forestry* 95, no. 5 (1997): 29–33.

Wondolleck, Julia M., and Clare M. Ryan. "What Hat Do I Wear Now? An Examination of Agency Roles in Collaborative Processes." *Negotiation Journal* 15 (1999): 117–133.

Wondolleck, Julia M., and Steven L. Yaffee. *Building Bridges Across Agency Boundaries: In Search of Excellence in the United States Forest Service*. Ann Arbor: University of Michigan School of Natural Resources and Environment, 1994.

Acknowledgments

The editors would like to thank all of the people and organizations that made this book possible. First, we would like to acknowledge the hard work of all of the contributors to this volume. Special thanks go to Bill Vaughn for his careful attention to detail in his design and composition work, and to Michael McCurdy for the beautiful woodblock prints. We would also like to recognize John Baden and his staff at Gallatin Writers, Inc., and the Liberty Fund for bringing the editors together for lively discussions that precipitated the formation of this book. We are also grateful for the many supporters of the *Chronicle of Community,* especially the William and Flora Hewlett Foundation, the Ford Foundation, the Henry P. Kendall Foundation, and the Liz Claiborne and Art Ortenberg Foundation. We also owe much to Bill Clarke and Lynanne Hawthorne of NLI, and to Dorothy O'Brien, who volunteered dozens of hours of typography to the *Chronicle of Community.* Special thanks go to Shirley Muse and Cora Heid at Whitman College for helping us with all of the real work of putting together a project like this one.

About the Contributors

PHILIP BRICK teaches international and environmental politics at Whitman College in Walla Walla, Washington. In 1996, he edited, with R. McGreggor Cawley, *A Wolf in the Garden: The Land Rights Movement and the New Environmental Debate*, published by Rowman and Littlefield.

BEVERLY BROWN is coordinator of the Jefferson Center for Education and Research, based in southern Oregon. She is author of *In Timber Country: Working People's Stories of Environmental Conflict and Urban Flight*, published by Temple University Press.

JIM BURCHFIELD is director of the Bolle Center for People and Forests at the University of Montana. Previously he worked for the U.S. Forest Service and the Peace Corps.

KELLY CASH is program director of The Nature Conservancy's Ranch Working Group, and has worked for the Conservancy since 1981. She currently lives in Reno, Nevada.

GEORGE CAMERON COGGINS is Tyler Professor of Law, University of Kansas School of Law. He is co-author, with Charles F. Wilkinson, of a leading textbook on public land law, as well as numerous law review articles.

CARLOTTA COLLETTE is a freelance writer and public involvement consultant living in Milwaukie, Oregon. Her areas of focus include fish and wildlife recovery, energy policy, efficiency, renewable resources, transportation, architecture, and agriculture. Collette was executive editor for the Northwest Power Planning Council from 1984 until 1998, where she published the council's prize-winning public information magazine.

ALEX CONLEY is a research assistant at the Udall Center for Studies in Public Policy in Tucson, Arizona.

MIKE CONNELLY is a hay farmer and cattle rancher near the town of Bonanza, Oregon, and is chair of the Cloverleaf Watershed Association.

SUSAN CULP is a Research Assistant for the Sonoran Institute, a nonprofit organization based in Tucson, Arizona. Culp's work focuses on community stewardship organizations (CSOs), nonprofit organizations created to balance conservation and development objectives, preserve local character and empower citizens to be good stewards of their natural, cultural, and economic resources. Culp received her B.A. in marine biology from the University of California at Santa Cruz, and is currently pursuing a master's degree in public administration and policy, with a concentration in natural resource policy, at the University of Arizona in Tucson.

PETER R. DECKER writes and ranches from Colorado and Nebraska. Once a college professor, a war correspondent in Vietnam for the Associated Press, and the Colorado Commissioner of Agriculture under Governor Roy Romer, Peter is the author of two books: *Fortunes and Failures*, a study of San Francisco's nineteenth-century merchants, and *Old Fences, New Neighbors*, an investigation into the transformation of the rural West.

DAVID H. GETCHES is the Raphael J. Moses Professor of Natural Resources Law, University of Colorado School of Law. He teaches and writes on water law, public lands, environmental policy, and Indian law and consults with governmental agencies and nongovernmental organizations in the United States and abroad. He was founding executive director of Native American Rights Fund and executive director of the Colorado Department of Natural Resources.

MALL JOHANI is an artist living in Chimacum, Washington.

LYNN JUNGWIRTH, of Hayfork, California, comes from generations of logging and sawmilling families. She directs the Watershed Center, a worker-based nonprofit dealing with the integration of healthy forests and healthy communities through research, training, and economic development. Ms. Jung-

WIRTH is the national chair of the Communities Committee of the Seventh American Forests Congress.

DANIEL KEMMIS is director of the O'Connor Center for the Rocky Mountain West, and former mayor of Missoula, Montana. He is author of *Community and the Politics of Place*, published by the University of Oklahoma Press, and *The Good City and the Good Life*, published by Houghton Mifflin.

DOUGLAS S. KENNEY holds a research faculty position at the Natural Resources Law Center, University of Colorado. His primary interest is institutional arrangements for the governance, administration, and use of natural resources. In recent years, he has overseen the center's research program on western watershed initiatives, examining key structural and functional qualities of these efforts for the benefit of policy makers, academics, field-level practitioners, and the general public.

JONATHAN I. LANGE is professor of communication at Southern Oregon University, where he teaches organization communication, communication theory, and negotiation and conflict management. For two years, he was cofacilitator for the Applegate Partnership.

ED MARSTON is publisher of *High Country News*.

MICHAEL MCCURDY has illustrated many award-winning books for children and adults. He lives with his wife and two children on a farm in Massachusetts.

MATTHEW J. MCKINNEY is director of the Montana Consensus Council. For the past fifteen years, he has been a student and practitioner of citizen involvement and dispute resolution in natural resource decisions.

CARL M. MOORE is Professor Emeritus at Kent State University. He co-chaired the gubernatorial commission that created the Ohio Office of Dispute Resolution and Conflict Management and has worked for many years, primarily with the Kettering Foundation, on strategies to help local officials use collaborative approaches.

CASSANDRA MOSELEY is an assistant professor in the Department of Political Science at University of Florida. Prior to coming to UF, she worked as program development director for the Rogue Institute for Ecology and Economy in Ashland, Oregon, and acted as its representative on the Applegate Partnership Board in 1996–97.

LUTHER PROPST is executive director of both the Sonoran Institute and the Rincon Institute, two affiliated, Tucson-based nonprofit organizations. Previously, Mr. Propst was a senior associate with World Wildlife Fund and The Conservation Foundation in Washington, D.C.

RAY RASKER is Director of the Sonoran Institute's Northwest office and spends much of his time helping rural communities understand recent changes in their economies, as well as the relation between environmental quality and community development.

WILLIAM E. RIEBSAME is associate professor of geography at the University of Colorado, Boulder. He studies land use and development patterns in the American West, and is general editor of *Atlas of the New West*, published by W. W. Norton in 1997.

SHIRLEY SOLOMON is chair of the Skagit Watershed Council in Washington State, and works with Long Live the Kings, as Seattle-based nonprofit dedicated to restoring wild salmon. A geographer by training, Shirley has specialized in all aspects of cooperative problem solving and collaborative processes relating to the land and its resources.

DONALD SNOW is executive editor of the *Chronicle of Community* and executive director of the Northern Lights Research and Education Institute in Missoula, Montana. He published *The Next West: Public Lands, Community, and Economy in the American West* with Island Press in 1997.

ANDY STAHL is executive director of Forest Service Employees for Environmental Ethics, a nationwide nonprofit conservation group. A forester, he has worked for the USDA-Forest Service, Associated Oregon Loggers, National Wildlife Federation, and Sierra Club Legal Defense Fund.

FRAN STEFAN is a freelance writer living in Seattle, with ten years' experience working with state, federal, and tribal governments.

EDWARD P. WEBER is assistant professor of political science at Washington State University. He is the author of *Pluralism by the Rules: Conflict and Cooperation in Environmental Regulation*, published by Georgetown University Press in 1998.

SARAH VAN DE WETERING created and was the original editor of the *Chronicle of Community*. She works as a writer/editor and policy consultant on western resource issues from her home in Missoula, Montana.

MARIA VARELA is a community organizer who has worked with rural communities since 1963. In 1990 she was awarded a MacArthur Foundation Fellowship for her life's work, including the founding of Ganados del Valle and Tierra Wools in northern New Mexico. Varela is on faculty at the Community and Regional Planning Department at the University of New Mexico. She is co-author of *Rural Environmental Planning for Sustainable Communities,* published in 1992 by Island Press. In 1997–98 Varela was awarded an endowed chair in Southwestern Studies Department at Colorado College and continues there as adjunct professor. Maria was named in 1998 to the "Bad Girl Hall of Fame" by *Ms.* magazine.

BILL VAUGHN is a writer and graphic designer in Missoula, Montana.

FLORENCE WILLIAMS has written about environmental issues and the West for ten years, first as a staff reporter at *High Country News.* Her articles and essays frequently appear in such publications as the *New York Times,* the *San Francisco Chronicle,* the *New Republic,* and *Outside* magazine. A graduate of the creative writing program at the University of Montana, she won the 1996 and 1998 personal essay awards from the American Society of Journalists and Authors. She currently lives in Helena, Montana, with her husband, Jamie Williams.

Index